Metal-Based Catalysts
in Organic Synthesis

Metal-Based Catalysts in Organic Synthesis

Editor

Manas Sutradhar

MDPI • Basel • Beijing • Wuhan • Barcelona • Belgrade • Manchester • Tokyo • Cluj • Tianjin

Editor
Manas Sutradhar
Centro de Química Estrutural,
Instituto Superior Técnico,
Universidade de Lisboa
Portugal

Editorial Office
MDPI
St. Alban-Anlage 66
4052 Basel, Switzerland

This is a reprint of articles from the Special Issue published online in the open access journal *Catalysts* (ISSN 2073-4344) (available at: https://www.mdpi.com/journal/catalysts/special_issues/metal_organic).

For citation purposes, cite each article independently as indicated on the article page online and as indicated below:

LastName, A.A.; LastName, B.B.; LastName, C.C. Article Title. *Journal Name* **Year**, *Volume Number*, Page Range.

ISBN 978-3-0365-1062-0 (Hbk)
ISBN 978-3-0365-1063-7 (PDF)

© 2021 by the authors. Articles in this book are Open Access and distributed under the Creative Commons Attribution (CC BY) license, which allows users to download, copy and build upon published articles, as long as the author and publisher are properly credited, which ensures maximum dissemination and a wider impact of our publications.

The book as a whole is distributed by MDPI under the terms and conditions of the Creative Commons license CC BY-NC-ND.

Contents

About the Editor vii

Manas Sutradhar
Metal-Based Catalysts in Organic Synthesis
Reprinted from: *Catalysts* **2020**, *10*, 1429, doi:10.3390/catal10121429 1

Tannistha Roy Barman, Manas Sutradhar, Elisabete C. B. A. Alegria, M. Fátima C. Guedes da Silva and Armando J. L. Pombeiro
Fe(III) Complexes in Cyclohexane Oxidation: Comparison of Catalytic Activities under Different Energy Stimuli
Reprinted from: *Catalysts* **2020**, *10*, 1175, doi:10.3390/catal10101175 3

Dominik Madej, Adrian Konopko, Piotr Piotrowski and Agnieszka Krogul-Sobczak
Pd Nanoparticles and Mixture of $CO_2/CO/O_2$ Applied in the Carbonylation of Aniline
Reprinted from: *Catalysts* **2020**, *10*, 877, doi:10.3390/catal10080877 19

Dandan Yang, Haiyan Wang, Wenhua Wang, Sihua Peng, Xiuzhen Yang, Xingliang Xu and Shouhua Jia
Nickel-Modified TS-1 Catalyzed the Ammoximation of Methyl Ethyl Ketone
Reprinted from: *Catalysts* **2019**, *9*, 1027, doi:10.3390/catal9121027 33

Rita N. Kadikova, Ilfir R. Ramazanov, Azat M. Gabdullin, Oleg S. Mozgovoi and Usein M. Dzhemilev
Carbozincation of Substituted 2-Alkynylamines, 1-Alkynylphosphines, 1-Alkynylphosphine Sulfides with Et_2Zn in the Presence of Catalytic System of Ti(O-*i*Pr)$_4$ and EtMgBr
Reprinted from: *Catalysts* **2019**, *9*, 1022, doi:10.3390/catal9121022 47

About the Editor

Manas Sutradhar received his Ph.D. in Chemistry from the University of Calcutta, India, in 2008. He was a postdoc fellow at the Johannes Gutenberg University of Mainz, Germany. Currently, he is a research scientist at the Instituto Superior Técnico, Universidade de Lisboa, Portugal. He has published more than 65 papers in international peer review journals (h-index 25). He has edited one book and published 7 book chapters in books with international circulation. The major area of his research is based on synthesis and catalysis. The use of microwave irradiation for oxidation catalysis, the chemistry of vanadium complexes and their applications are other important areas of his interest.

Editorial
Metal-Based Catalysts in Organic Synthesis

Manas Sutradhar

Centro de Química Estrutural, Instituto Superior Técnico, Universidade de Lisboa, Av. Rovisco Pais, 1049-001 Lisboa, Portugal; manas@tecnico.ulisboa.pt

Received: 30 November 2020; Accepted: 3 December 2020; Published: 7 December 2020

The role of catalysts is extremely important for various organic transformations and the synthesis of organic compounds. In the last two decades, a significant growth has been observed in the areas of catalytic processes with the development of several catalytic methods e.g., transition metal catalysis, organocatalysis, photocatalysis, electrocatalysis and biocatalysis [1,2]. Metal based catalysts are a widely used catalytic system and have notable contribution towards the synthesis of organic compounds. They can be simple metal salts, metal oxides or metal complexes which can exhibit high catalytic activity. They have diverse applications as catalysts under homogeneous, heterogenous or onto solid support conditions, from academic to industrial research laboratories for the transformation of organic compounds [3]. Several techniques have already been explored, such as the use of microwave irradiation, ultrasound, light assisted or the use of ionic liquid or a supercritical CO_2 medium to drive the catalytic reaction energy efficiently. The development of new, sustainable and energy efficient catalytic processes is a great challenge for the synthesis of organic compounds.

This special issue is dedicated to contemporary progress of new metal-based catalytic systems for the synthesis of organic compounds. Four research articles address four different areas of metal-based catalytic processes towards the development of organic synthesis.

The first article deals with the catalytic carbozincation of 2-alkynylamines, 1-alkynylphosphine and -phosphine sulfides [4].

The EtMgBr and Ti(O-iPr)$_4$-catalyzed reaction of the mentioned compounds with Et$_2$Zn has been studied in various solvents, such as diethyl ether, methylene chloride, hexane, toluene, benzene, and anisole. Selective formations of 2-alkenylamines and 1-alkenylphosphine oxides from 2-alkynylamines and 1-alkynylphosphines are obtained after oxidation with H_2O_2 in diethyl ether, whereas a mixture of stereoisomers is formed in other solvents. In 1989, Kulinkovich et al. first reported the formation of a titanacyclopropane complex upon the reaction of ethylmagnesium bromide with titanium (IV) alkoxides [5]. In this work, the reaction of (cyclopropylethynyl) diphenylphosphine sulfide was non-regioselective and produced a mixture of 1-alkenylphosphines regioisomers. This study shows a useful pathway for the transformations of functionalized acetylene derivatives using metal complex catalyzed organozinc synthesis. A proposed mechanism of Ti-Mg-catalyzed carbozincation of substituted 1-alkynylphosphine sulfides with Et$_2$Zn is also illustrated.

The second article describes the catalytic performances of a series of transition metal (Fe, Co, Ni, Cu, Ce)-modified titanium silicalite-1 (M-TS-1) catalysts towards the ammoximation of methyl ethyl ketone (MEK) [6]. Ultrasonic impregnation method was applied for the preparation of transition metal-modified TS-1 [7]. The nickel-modified TS-1 catalyst exhibits good conversion and high selectivity of methyl ethyl ketoxime (MEKO) with H_2O_2 as the oxidant. It also shows efficient recyclable properties [6]. The nickel modification changes the electron environment of the Ti active site and optimizes the adsorption capacity for H_2O_2 activation. It also helps to reduce the surface acidity of the catalyst and prevents the further oxidation of oxime.

The third article reports the use of Pd nanoparticles and a mixture of $CO_2/CO/O_2$ towards the carbonylation of aniline to N,N'-diphenylurea [8]. PdCl$_2$(2,4Cl$_2$Py)$_2$, PdCl$_2$(4MePy)$_2$, PdCl$_2$, palladium nanoparticles (PdNPs), and palladium-based nanostructural material (PdNM) are studied as pre-catalyst for the carbonylation of aniline. The highest conversion and selectivity of aniline to N,N'-diphenylurea is reported for PdNPs with the mixture of CO/O_2 and CO_2 as the carbonylating agent. The proposed catalytic cycle shows that Pd(0) stabilized by pyridine ligand is the active species for the carbonylation reaction. This hypothesis attributes the efficiency of PdNPs over Pd(II) complexes.

The last article by Roy Barman et al. elucidates the comparison of catalytic activities towards the oxidation of cyclohexane with Fe(III) complexes under different energy stimuli (microwave irradiation, ultrasound and conventional heating) [9].

The oxidation of cyclohexane to cyclohexanol and cyclohexanone has a high industrial significance for the manufacture of Nylon-6,6 [10]. Three different Fe(III) compounds derived from N'-acetylpyrazine-2-carbohydrazide (H$_2$L) have been reported and their catalytic activities are compared in three different energy stimuli. The iron complex [Fe(HL)(NO$_3$)(H$_2$O)$_2$]NO$_3$ (**1**) shows the highest catalytic activity under optimized microwave irradiation conditions compared to [Fe(HL)Cl$_2$] (**2**) and [Fe(HL)Cl(μ-OMe)]$_2$ (**3**) [9].

This special issue is versatile in terms of catalytic reactions and their applications. This collection will bring much interest to the researchers in the field of catalysis.

I would like to express my sincere gratitude to all authors for their valuable contributions to make this issue successful.

Conflicts of Interest: The authors declare no conflict of interest.

References

1. Beller, M.; Renken, A.; van Santen, R.A. (Eds.) *Catalysis: From Principles to Applications*, 1st ed.; Wiley-VCH Verlag GmbH & Co. KGaA: Weinheim, Germany, 2012.
2. Sutradhar, M.; da Silva, J.A.L.; Pombeiro, A.J.L. (Eds.) *Vanadium Catalysis*; Royal Society of Chemistry: Cambridge, UK, 2020. [CrossRef]
3. Knochel, P.; Molander, G.A. (Eds.) *Comprehensive Organic Synthesis*; Elsevier: Amsterdam, The Netherlands, 2014.
4. Kadikova, R.N.; Ramazanov, I.R.; Gabdullin, A.M.; Mozgovoi, O.S.; Dzhemilev, U.M. Carbozincation of Substituted 2-Alkynylamines, 1-Alkynylphosphines, 1-Alkynylphosphine Sulfides with Et$_2$Zn in the Presence of Catalytic System of Ti(O-iPr)$_4$ and EtMgBr. *Catalysts* **2019**, *9*, 1022. [CrossRef]
5. Kulinkovich, O.G.; Sviridov, S.V.; Vasilevskii, D.A.; Pritytskaya, T.S. Reaction of ethylmagnesium bromide with carboxylic esters in the presence of tetraisopropoxytitanium. *Zh. Org. Khim.* **1989**, *25*, 2244–2245.
6. Yang, D.; Wang, H.; Wang, W.; Peng, S.; Yang, X.; Xu, X.; Jia, S. Nickel-Modified TS-1 Catalyzed the Ammoximation of Methyl Ethyl Ketone. *Catalysts* **2019**, *9*, 1027. [CrossRef]
7. Wu, M.; Chou, L.; Song, H. Effect of metals on titanium silicalite TS-1 for butadiene epoxidation. *Chin. J. Catal.* **2013**, *34*, 789–797. [CrossRef]
8. Madej, D.; Konopko, A.; Piotrowski, P.; Krogul-Sobczak, A. Pd Nanoparticles and Mixture of $CO_2/CO/O_2$ Applied in the Carbonylation of Aniline. *Catalysts* **2020**, *10*, 877. [CrossRef]
9. Roy Barman, T.; Sutradhar, M.; Alegria, E.C.B.A.; Guedes da Silva, M.F.C.; Pombeiro, A.J.L. Fe(III) Complexes in Cyclohexane Oxidation: Comparison of Catalytic Activities under Different Energy Stimuli. *Catalysts* **2020**, *10*, 1175. [CrossRef]
10. Pombeiro, A.J.L. Chapter 1: Alkane Functionalization: Introduction and overview. In *Alkane Functionalization*; Pombeiro, A.J.L., Guedes da Silva, M.F.C., Eds.; Wiley: Hoboken, NJ, USA, 2019; pp. 1–15.

Publisher's Note: MDPI stays neutral with regard to jurisdictional claims in published maps and institutional affiliations.

© 2020 by the author. Licensee MDPI, Basel, Switzerland. This article is an open access article distributed under the terms and conditions of the Creative Commons Attribution (CC BY) license (http://creativecommons.org/licenses/by/4.0/).

Article

Fe(III) Complexes in Cyclohexane Oxidation: Comparison of Catalytic Activities under Different Energy Stimuli

Tannistha Roy Barman [1], Manas Sutradhar [1,*], Elisabete C. B. A. Alegria [1,2,*], Maria de Fátima C. Guedes da Silva [1] and Armando J. L. Pombeiro [1,*]

[1] Centro de Química Estrutural, Instituto Superior Técnico, Universidade de Lisboa, Av. Rovisco Pais, 1049-001 Lisboa, Portugal; roybarman@tecnico.ulisboa.pt (T.R.B.); fatima.guedes@tecnico.ulisboa.pt (M.d.F.C.G.d.S.)
[2] Departamento de Engenharia Química, Instituto Superior de Engenharia de Lisboa, Instituto Politécnico de Lisboa, R. Conselheiro Emídio Navarro, 1, 1959-007 Lisboa, Portugal
* Correspondence: manas@tecnico.ulisboa.pt (M.S.); elisabete.alegria@isel.pt (E.C.B.A.A.); pombeiro@tecnico.ulisboa.pt (A.J.L.P.)

Received: 25 September 2020; Accepted: 9 October 2020; Published: 13 October 2020

Abstract: In this study, the mononuclear Fe(III) complex [Fe(HL)(NO$_3$)(H$_2$O)$_2$]NO$_3$ (**1**) derived from N'-acetylpyrazine-2-carbohydrazide (H$_2$L) was synthesized and characterized by several physicochemical methods, e.g., elemental analysis, infrared (IR) spectroscopy, electrospray ionization mass spectrometry (ESI-MS), and single crystal X-ray diffraction analysis. The catalytic performances of **1** and the previously reported complexes [Fe(HL)Cl$_2$] (**2**) and [Fe(HL)Cl(μ-OMe)]$_2$ (**3**) towards the peroxidative oxidation of cyclohexane under three different energy stimuli (microwave irradiation, ultrasound, and conventional heating) were compared. **1-3** displayed homogeneous catalytic activity, leading to the formation of cyclohexanol and cyclohexanone as final products, with a high selectivity for the alcohol (up to 95%). Complex **1** exhibited the highest catalytic activity, with a total product yield of 38% (cyclohexanol + cyclohexanone) under optimized microwave-assisted conditions.

Keywords: Fe(III) complex; N,O donor; X-ray analysis; alkane oxidation; microwave irradiation

---

1. Introduction

The functionalization of alkanes should provide convenient methods for building a range of valuable organic products [1–9]. However, due to their inertness, such processes are difficult and, e.g., the catalytic oxidation of alkanes remains one of the most challenging fields of chemistry.

The selective oxidation of cyclohexane to cyclohexanol and cyclohexanone has a significant relevance in terms of industrial and economical viewpoints, because these products are used as precursors of adipic acid which, among other applications, is used in the manufacturing of Nylon-6,6 [1–14]. In industry, that reaction is currently performed using a cobalt(II) naphthenate catalyst [15] at 160 °C and 15 bar with very low yields (4%), in order to achieve a good selectivity (80%) [1,2,9,14,15]. The search for more efficient catalytic systems through the development of novel metal-based catalysts and single-pot methodologies for the mild oxidation of alkanes is thus a challenging task [1–3,7,16–23], involving new catalysts with cheap and abundant metals, namely copper [24–27] and iron [28–33]. Microwave irradiation has also been shown to assist with the reaction in various cases [28–33].

In continuation of our work on the catalytic oxidation of alkanes and related species with metal-hydrazone catalysts [24–27,34–44], herein, we report the synthesis and characterization of the new mononuclear Fe(III) complex [Fe(HL)(NO$_3$)(H$_2$O)$_2$]NO$_3$ (**1**) obtained from N'-acetylpyrazine-2-carbohydrazide (H_2L), and a comparison of its catalytic activity with those of the related compounds [Fe(HL)Cl$_2$] (**2**) and [Fe(HL)Cl(μ-OMe)]$_2$ (**3**) [45]. The catalytic activity of these complexes in the oxidation of cyclohexane was evaluated by conventional heating and under microwave (MW) or ultrasound (US) irradiations. MW and US irradiations were applied to activate the proposed catalytic system, eventually enhancing the yield, selectivity, and/or rate in comparison with conventional heating [46,47], but their application in alkane functionalization is still rather underexplored [48–50].

2. Results and Discussion

The reaction between Fe(NO$_3$)$_3$·9H$_2$O and the pro-ligand N'-acetylpyrazine-2-carbohydrazide (H$_2$L) in methanol results in the formation of the mononuclear Fe(III) complex [Fe(HL)(NO$_3$)(H$_2$O)$_2$]NO$_3$ (**1**) (Scheme 1). Complexes **2** and **3** were synthesized according to the literature [45]. These complexes were characterized by elemental analysis, infrared (IR) spectroscopy, and electrospray ionization mass spectrometry (ESI-MS) and, for **1**, also by X-ray single crystal diffraction analysis. The expected characteristic stretching bands of HL$^-$ in **1** are shifted to lower wavenumbers relative to those in the pro-ligand (see the experimental section), specifically the ν(NH) at 3136 cm^{-1} and the ν(C=O) detected at 1678 and 1646 cm^{-1}; the band for nitrate ions occurs at the typical value of 1383 cm^{-1} [51]. The molecular ion was not observed in the ESI-MS analysis (see the experimental section), but instead, the species [M-(NO$_3$)]$^+$ resulting from the loss of the nitrate counter-ion was detected.

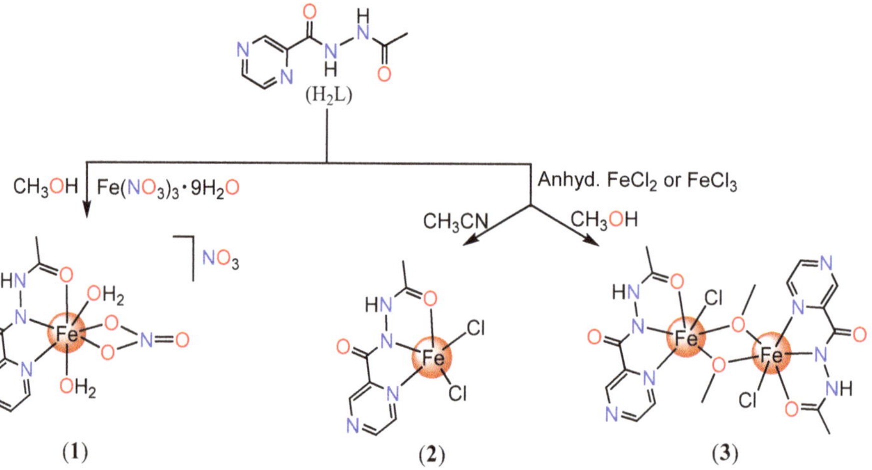

Scheme 1. Syntheses of **1**–**3**.

2.1. Crystal Structure

Crystals of **1** (Figure 1) were obtained from methanol upon slow evaporation, at room temperature. A summary of the crystallographic data and processing parameters is presented in Table 1. Complex **1** (Supplementary Materials: CCDC number 2019464) crystallized in the monoclinic $P2_1/c$ space group, with the asymmetric unit containing an iron cation with N'-acetylpyrazine-2-carbohydrazide acting as a mononegative $N_{pyrazine}N_{amido}O_{keto}$ chelate ligand and a OO'-donor nitrate anion sharing the equatorial binding region. The axial sites were engaged with two water ligands, and a cationic pentagonal bipyramidal iron complex was thus formed, with its charge being balanced by a nitrate counter anion. The unit cell of **1** contains five non-coordinated and disordered water molecules. The $NN'O$ coordination mode of HL$^-$ was observed in other cases [45,51] and differs from that found in complexes with ligands derived from N-acetylsalicylhydrazide [34,52,53]. The HL$^-$ ligand in **1** is slightly twisted, as measured by the distance between the least-square plane defined by the pyrazine ring and the C_{methyl} atom (0.539 Å); in **3**, that distance is shorter (0.388 Å) and in **2**, such moieties are coplanar [45]. Due to the water ligands and non-coordinated water molecules, compound **1** is involved in extensive H-bond interactions, which extend the structure to the third dimension. This contrasts with what was found in **2** and **3**, resulting from the kind of contact [45], with the former being assembled in dimers and the latter giving rise to 1D chains.

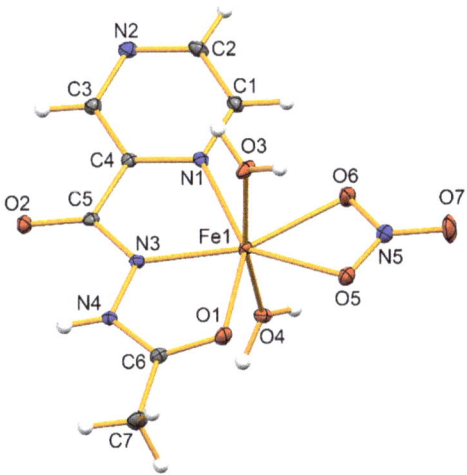

Figure 1. Ellipsoid plot of the complex cation in **1** (drawn at a 30% probability level) with an atom labeling scheme. Selected bond distances (Å) and angles (°): N3–N4 1.389(3); C5–O2 1.241(3); C6–O1 1.266(3); N1–Fe1 2.2547(19); N3–Fe1 2.034(2); O1–Fe1 2.0707(17); O3–Fe1 1.9998(17); O4–Fe1 1.9910(17); O5–Fe1 2.1163(17); O6–Fe1 2.2508(18); O4–Fe1–O3 167.97(8); O1–Fe1–N1 146.91(7); N3–Fe1–N1 72.68(7); N3–Fe1–O1 74.29(7); and O3–Fe1–N3 95.88(7).

Table 1. Crystallographic parameters and structure refinement data for complex **1**.

	1
Empirical formula	$C_7H_{11}FeN_6O_{11.25}$
Formula weight	415.04
Crystal system	Monoclinic
Space group	$P2_1/c$
Temperature/K	150(2)
a/Å	7.2541(4)
b/Å	14.4633(8)
c/Å	14.7415(9)
β/°	93.275(2)
V (Å^3)	1544.13(15)
Z	4
D_{calc} (g cm^{-3})	1.785
F000	844
µ(Mo Kα) (mm^{-1})	1.052
Rfls. collected/unique/observed	22957/2837/2385
R_{int}	0.0478
Final $R1^a$, $wR2^b$ ($I \geq 2\sigma$)	0.0319, 0.0711
Goodness-of-fit on F^2	1.051

[a] $R = \Sigma||F_o|-|F_c||/\Sigma|F_o|$; [b] $wR(F^2) = [\Sigma w(|F_o|^2 - |F_c|^2)^2/\Sigma w|F_o|^4]^{\frac{1}{2}}$.

The catalytic properties of **1-3** were investigated and compared for the peroxidative oxidation of cyclohexane, and the effect of the type of activating energy input on the catalytic output was also studied.

2.2. Peroxidative Oxidation of Cyclohexane

Using aqueous *tert*-butylhydroperoxide (*t*-BuOOH, 70% aqueous solution) as an oxidant in acetonitrile (NCMe) at 50 °C, the homogeneous **1-3** were screened for the microwave-assisted (MW) oxidation of cyclohexane (CyH) to cyclohexanone (Cy=O) and cyclohexanol (CyOH) as the final products (Scheme 2). The detection of no other products by Gas Chromatography–Mass Spectrometry (GC-MS) analysis suggests that the catalytic oxidation is very selective. The mechanism involved in the cyclohexane oxidation, involving the generation of cyclohexyl hydroperoxide (CyOOH) as a primary product, was corroborated by the method proposed by Shul'pin [54] and the reaction mixture injected in the gas chromatograph before and after treatment with triphenylphosphine (PPh$_3$). The amount of CyOH had increased significantly after the addition of PPh$_3$ to the reaction mixture, due to the reduction of CyOOH to CyOH and formation of phosphine oxide (PPh$_3$O) (Scheme 2). The results are reported in Table 2 (yield values refer to the samples after treatment with an excess of PPh$_3$).

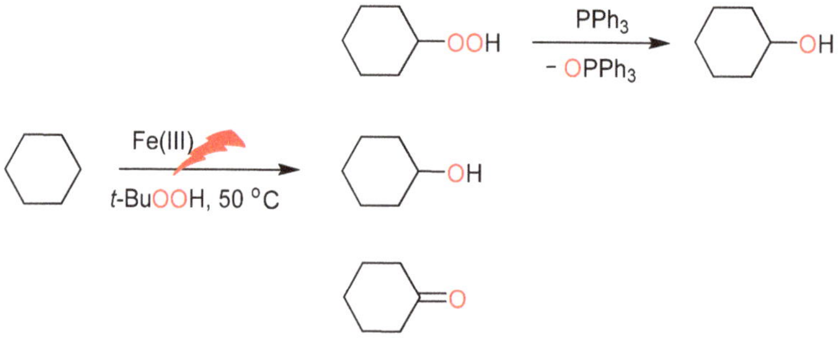

Scheme 2. Microwave (MW)-assisted oxidation of cyclohexane (CyH).

In the presence of **1** and after 3 h under MW-irradiation at 50 °C, 14.9% of cyclohexane was converted into ketone-alcohol (KA) oil (cyclohexanol and cyclohexanone mixture), with cyclohexanol as the major product (selectivity up to 86% relative to KA oil after treatment with PPh$_3$), in the absence of any co-catalyst (entry 5, Table 2), with a turnover number of 75 per Fe atom. There was always an increase in the amount of alcohol when the reaction mixture was analyzed after the addition of triphenylphosphine with respect to the existing amount before this treatment. This fact suggests the formation of cyclohexyl hydroperoxide (CyOOH) as a primary product which, after treatment with PPh$_3$, is reduced to cyclohexanol (CyOH). As an example, entries 4 and 5 from Table 2 are presented, and from an initial CyOH yield of 7.5% (before PPh$_3$ treatment), an increment to 13.2% was detected (after PPh$_3$ treatment). Furthermore, a decrease in the amount of cyclohexanone (Cy=O) was observed, although with a less pronounced difference. For iron complexes **2** and **3**, and for the same reaction conditions, total yields of 10.4% and 14.4% were obtained, with selectivities of 88% and 85% for the alcohol, respectively (Table 2, entries 22 and 25).

The oxidation reaction of cyclohexane was attempted under solvent-free conditions (in the absence of acetonitrile) in the presence of **1** (Table 2, entry 3), without success (the yield did not go beyond 0.74%, despite the high selectivity for the alcohol of 95%). The replacement of compounds **1-3** by their precursor salts, Fe(NO$_3$)$_3$·9H$_2$O and FeCl$_2$, resulted in much lower conversions, with total yields not exceeding 5% (Table 2, entries 28 and 29, respectively), denoting the significance of the NN'O-donor ligand N'-acetylpyrazine-2-carbohydrazide (HL$^-$) in the Fe coordination sphere in the promotion of the catalytic performance of complexes **1-3**. The free pro-ligand did not exhibit any activity. Blank tests were performed in the absence of any of the Fe(III) compounds **1-3**, but no noteworthy conversion was observed.

Table 2. Data [a] for the MW-assisted oxidation of cyclohexane by catalysts **1-3** using TBHP (70% aqueous) as an oxidant.

Entry	Catalyst	Catalyst Amount (mol)	Reaction Time (h)	Yield (%) [b] CyOH	Cy=O	TOTAL [c]	Selectivity to Cyclohexanol (%) [d]
1		10	0.5	1.6	0.1	1.7	94
2		10	1	6.7	0.7	7.4	91
3 [e]		10	1	0.7	0.04	0.74	95
4 [f]		10	3	7.5	3.1	10.6	71
5		10	3	13.2	1.7	14.9	86
6		10	6	14.2	2.7	16.9	84
7 [g]		10	3	7.6	1.1	8.7	87
8 [h]		10	3	16.7	4.8	21.5	78
9		2.5	3	3.6	0.2	3.8	95
10		5	3	8.4	0.9	9.3	90
11	1	20	3	15.7	3.7	19.4	80
12 [i]		10	3	17.1	4.3	21.4	80
13 [j]		10	3	22.6	6.1	28.7	79
14 [k]		10	3	24.8	12.9	37.7	66
15 [l]		10	3	27.5	9.7	37.2	74
16 [k]		10	1	9.9	1.7	11.6	85
17 [k]		10	6	24.7	12.9	37.6	66
18 [k]		10	9	23.5	13.6	37.1	63
19 [m]		10	3	7.7	1.1	8.8	88
20 [n]		10	3	21.8	6.5	28.3	77
21 [o]		10	3	1.3	0.4	1.7	76
22		10	3	9.1	1.3	10.4	88
23 [k]	2	10	3	9.8	0.9	10.7	92
24 [o]		10	3	2.5	0.2	2.7	93
25		10	3	12.3	2.1	14.4	85
26 [k]	3	10	3	16.3	1.9	18.2	90
27 [o]		10	3	1.9	0.4	2.3	83
28	Fe(NO$_3$)$_3$·9H$_2$O	10	3	3.8	1.5	5.3	72
29	Anhy. FeCl$_2$	10	3	3.3	0.9	4.2	79

[a] Reaction conditions (unless stated otherwise): cyclohexane (5.0 mmol); 2.5-20 μmol of catalyst; acetonitrile (3 mL); TBHP 70% aqueous solution (10 mmol); 1-9 h; 50 °C; microwave irradiation (5 W); yield and Turnover number (TON) determined by gas chromatography upon treatment with PPh$_3$ (see text). [b] Molar yield (%) based on substrate, i.e., moles of product [cyclohexanol (CyOH) or cyclohexanone (Cy=O)] per 100 moles of cyclohexane after PPh$_3$ treatment. [c] Total yield = moles of products [cyclohexanol (CyOH) + cyclohexanone (Cy=O)]/100 moles of cyclohexane. [d] Selectivity to cyclohexanol relative to KA oil mixture, i.e., moles of CyOH/(100 moles of CyOH + Cy=O). [e] Solvent-free conditions. [f] Without PPh$_3$ treatment. [g] Oxidant:substrate = 1:1. [h] Oxidant:substrate = 4:1. [i] n(HNO$_3$)/n(catalyst) = 5. [j] n(HNO$_3$)/n(catalyst) = 10. [k] n(HNO$_3$)/n(catalyst) = 25. [l] n(HNO$_3$)/n(catalyst) = 50. [m] n(HPCA)/n(catalyst) = 25. [n] n(TFA)/n(catalyst) = 25. [o] n(TEMPO)/n(catalyst) = 25.

Complex **1** was chosen for optimizing the reaction conditions and several parameters were explored. The effect of the reaction time is shown in Figure 2a, for the period between 0.5 and 6 h, without any additive and at 50 °C (Table 2, entries 1, 2, 5, and 6). There was a gradual increase over time in the amount of KA oil formed, achieving a total yield of ca. 15 and 17% after 3 and 6 h, respectively. During the first hour, the formation of cyclohexanone was not significant (94 and 91% of selectivity for the alcohol, after 0.5 and 1 h reaction, respectively). The yield of cyclohexanone increased (from 0.1 to 1.7%, after 0.5 and 3 h, respectively) and the selectively to the alcohol decreased to 86%, possibly due to the partial oxidation of cyclohexanol.

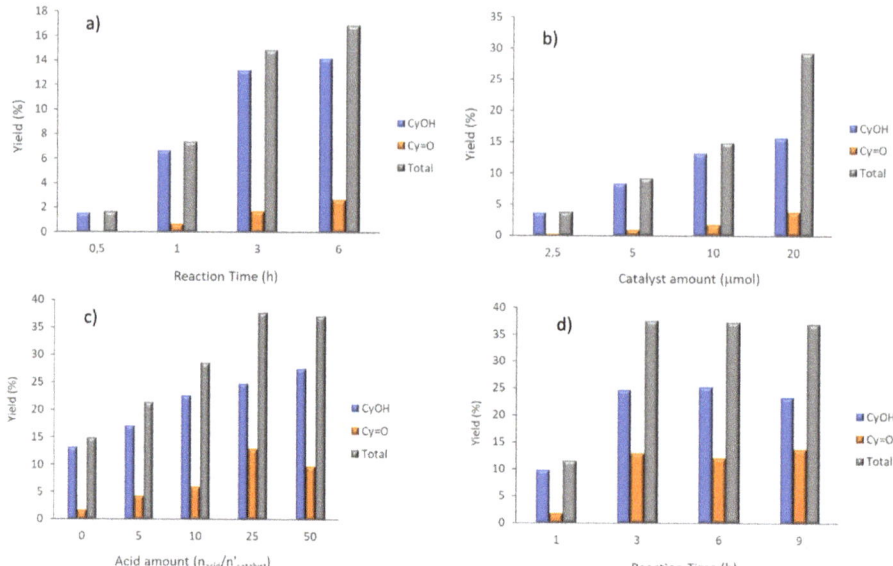

Figure 2. Oxygenated product yield of cyclohexane (CyH) oxidation [cyclohexanol (CyOH) + cyclohexanone (Cy=O)] with respect to the following: (**a**) reaction time (10 μmol of **1**); (**b**) catalyst amount (3 h reaction time); (**c**) acid amount (10 μmol of **1** and 3 h reaction time); and (**d**) reaction time in the presence of acid additive [10 μmol of **1** and n(HNO$_3$)/n'(catalyst) = 25]. Other reaction conditions: CyH (5 mmol); NCMe (3 mL); TBHP (70% aqueous) (10 mmol); 50 °C; under MW-irradiation (5 W); and GC analysis after the addition of PPh$_3$.

The amount of catalyst varied between 2.5 and 20 μmol (Figure 2b), and a maximum yield of 19.4% of KA oil was verified for an amount of 20 μmol of **1** (entry 11, Table 2). This parameter does not affect only the yield, which increases from 3.8 to 19.4% with the indicated range amount of catalyst, but also the selectivity which, for cyclohexanol, decreases from 95 to 80% (entries 1 and 9-1, Table 2). In this way, we can conclude that both the increase of the reaction time and the increase in the amount of catalyst favor the formation of ketone.

When performing the oxidation reactions under the typical conditions (3 h, 50 °C, and 10 μmol of **1**), the variation in the oxidant/substrate molar ratio also has an important effect on the oxidized product yields (Table 2, entries 5, 7, and 8). The yield of KA oil increased from 8.7 to 14.9 and afterwards to 21.5% when the oxidant/substrate molar ratio was changed from 1:1 to 2:1 and then to 4:1, respectively.

The effect of various additives on the peroxidative microwave-assisted oxidation of cyclohexane was also investigated. In the presence of nitric acid (HNO$_3$), complex **1** exhibited a highly promising effect. The n(HNO$_3$)/n'(catalyst) molar ratio changed from 5 to 50 (Figure 2c). The total yield increased to 37.7%, for n(HNO$_3$)/n'(catalyst **1**) = 25, relative to 14.9% obtained in the absence of any additive (Table 2, entries 5 and 14, respectively). Going beyond this molar ratio does not lead to a significant yield change (37.2% for n/n' = 50) (Table 2, entries 14 and 15).

To investigate the effect of the acid additive over time, catalytic oxidation was performed in the presence of this acid additive (n/n' = 25) for several time periods (1, 3, 6, and 9 h). Firstly, for both cases, in the absence and presence of HNO$_3$ (n/n' = 25) (Figure 2a,d), the quantity of oxygenated products (CyOH + Cy=O) practically reached the maximum after 3 h and then seemed to stabilize.

The positive effect of nitric acid has been observed for other catalytic systems involving the oxidative transformation of alkanes [26,27,55–57]. The presence of a certain amount of acid can promote the catalytic process, either by catalyst activation through the protonation of ligands and unsaturation of the metal center, or by promoting the properties of the oxidant.

The presence of HNO_3 affects, apart from the total yield, the product distribution. In the case of **1**, the selectivity for the alcohol was lower in the presence of acid, with the prolongation of time accentuating this effect. In the presence of acid, there was a clear preferential formation of CyOH (selectivity of 85%) in the first hour; a decrease of CyOH selectivity accompanied by an increase of Cy=O selectivity in the 1-3 h period; and beyond 3 h, the ratio between both products seemed to stabilize (Figure 2d).

In the presence of **1**, the influences of the 2-pyrazine carboxylic acid (HPCA), trifluroacetic acid (TFA), and stable free radical 2,2,6,6-tetramethylpiperidin-l-oxyl (TEMPO) were also explored (Figure 3). After 3 h at 50 °C under MW-irradiation, the total yield of products dropped in the presence of HPCA (8.8%) and TEMPO (drastically to 1.7%), whereas in the presence of TFA, it increased from 14.9% to 28.3% (Table 2, entries 14, 19, 20, and 21), although not as effectively as for HNO_3.

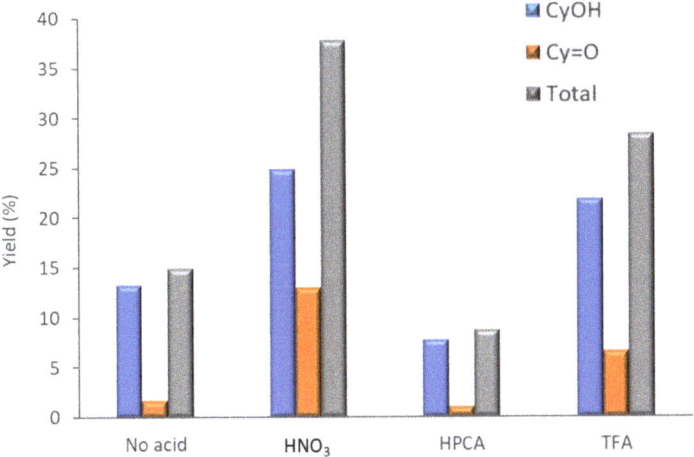

Figure 3. Effects of different additives on the oxidation of cyclohexane catalyzed by **1**. Reaction conditions: CyH (5 mmol); **1** (10 μmol); TBHP (70% aq.) (10 mmol); NCMe (3 mL); 50 °C; 3 h; under MW-irradiation (5 W).

The effects of the presence of HNO_3 (n(HNO_3)/n'(catalyst **2** or **3**) = 25) and TEMPO additives were also analyzed for compounds **2** and **3**. In the case of the acid additive, although an increase in the amount of oxygenated products was observed, this effect was not so accentuated for these catalysts as for **1** (Table 2, entries 23 and 26, for **2** and **3**, respectively). The presence of the TEMPO radical resulted, for both catalytic systems, as in the case of **1**, in a drastic decrease in the yields (Table 2, entries 24 and 27, for **2** and **3**, respectively).

The peroxidative oxidation of cyclohexane was also performed, for comparative purposes, using different types of energy inputs, apart from microwaves, namely conventional heating and ultrasound (US) irradiation. Reactions were performed for compounds **1-3**, during 3 h at 50 °C and in the presence of HNO_3 as an additive (n(HNO_3)/n'(catalyst) = 25), and in the case of compound **1**, for different periods of time.

If we consider the period of 3 h, compounds **1-3** responded differently to the different energy stimuli. It can be observed that **1** exhibits a better performance when the reaction is promoted by microwave radiation (Table 3, entry 8), conceivably due to its ionic character and larger dipole, which promote microwave energy absorption [31,48,58–60]. Accordingly, the effect of MW irradiation (in comparison with conventional heating) in the case of catalyst **3**, with a symmetrical apolar molecule, is negligible (Table 3, entries 17 and 18). In this case, a different driving force, i.e., acoustic cavitation in sonochemistry, shows a more effective role (Table 3, entry 19), which is consistent with the known ultrasonic cleavage of a metal-ligand bond [61]. For the dinuclear catalyst **3**, this can lead to the formation of more active mononuclear catalytic species. The reaction catalyzed by **2** does not seem to be favored by any of the radiations (MW and US), which may be due to its possible decomposition into less active or inactive species when it is under these energy inputs. It is also noteworthy to mention that, for the same period, the selectivity does not vary much when we compare the catalytic activity of **1-3** under the effect of different energy inputs (Table 3). For example, for the period of 3 h, and in the presence of **1**, the selectivity for cyclohexanol varies between 66 and 69% for different energy sources. In the case of **2** and **3**, only the MW stands out and the selectivities reach values higher than 90%.

Table 3. Effect of different energy inputs for the oxidation of cyclohexane by catalysts **1-3** using TBHP (70% aq.) as an oxidant [a].

Entry	Catalyst	Method	Reaction Time (h)	Yield (%) [b]			Selectivity to Cyclohexanol (%) [d]
				CyOH	Cy=O	TOTAL [c]	
1			0.5	16.5	5.1	21.6	76
2			1	16.9	6.2	23.1	73
3		CONV	3	18.9	8.6	27.5	69
4			6	19.8	7.8	27.6	72
5			24	28.8	11.8	40.6	71
6	1		0.5	11.6	2.9	14.5	82
7 [e]			1	15.2	4.6	19.8	77
8 [e]		MW	3	24.8	12.9	37.7	66
9 [e]			6	24.7	12.9	37.6	66
10			0.5	16.4	4.9	21.3	77
11		US	1	16.2	6.5	22.7	71
12			3	15.4	7.7	23.1	67
13			6	14.9	9.7	24.6	61
14		CONV	3	15.2	6.9	22.1	69
15	2	MW	3	9.8	0.9	10.7	92
16		US	3	7.3	4.0	11.3	65
17		CONV	3	13.3	4.4	17.7	75
18	3	MW	3	16.3	1.9	18.2	90
19		US	3	20.9	9.1	29.9	70

[a] Reaction conditions (unless stated otherwise): cyclohexane (5.0 mmol); 10 μmol of catalyst; acetonitrile (3 mL); TBHP 70% aqueous solution (10 mmol); n(HNO$_3$)/n′(catalyst) = 25; 0.5-6 h; 50 °C; microwave irradiation (5 W); yield and TON determined by gas chromatography upon treatment with PPh$_3$. [b] Molar yield (%) based on substrate, i.e., moles of product [cyclohexanol (CyOH) or cyclohexanone (Cy=O)] per 100 moles of cyclohexane after PPh$_3$ treatment. [c] Total yield = moles of products [cyclohexanol (CyOH) + cyclohexanone (Cy=O)]/100 moles of cyclohexane. [d] Selectivity to cyclohexanol relative to KA oil mixture, i.e., moles of CyOH/(100 moles of CyOH + Cy=O). [e] Data from Table 2 (entries 14, 16, and 17). CONV = under conventional heating with oil bath; MW = under MW-assisted condition; US = under ultrasound irradiation.

In addition, compound **1** was exposed to different types of energy input for different reaction times and, in all of the cases, a maximum selectivity for cyclohexanol was observed in the first 30 min of the reaction (Figure 4, Table 3). Thereafter, the selectivity decreased, conceivably due to the conversion of the cyclohexanol into cyclohexanone.

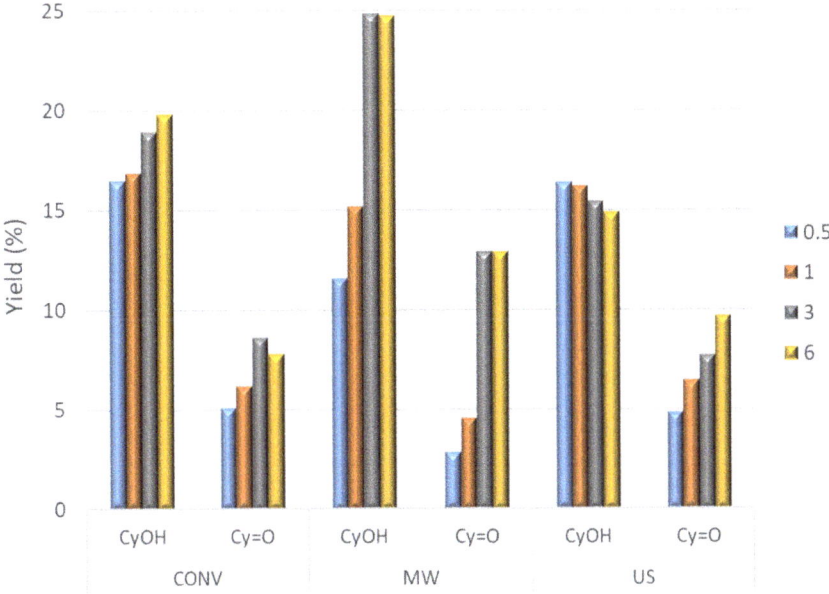

Figure 4. Effect of different energy stimuli on the oxidation of cyclohexane catalyzed by **1**. Reaction conditions: CyH (5 mmol); **1** (10 μmol); TBHP (70% aqueous) (10 mmol); NCMe (3 mL); n(HNO$_3$)/n'(catalyst) = 25; 50 °C; 0.5-6 h. CONV = under conventional heating with oil bath; MW = under MW-assisted condition; US = under ultrasound irradiation.

The present Fe(III) catalytic system is more effective for the oxidation of cyclohexane under MW conditions in terms of yields, compared to Cu(II) or V(V) catalytic systems with hydrazone-based ligands [27,35,42]. In the presence of an additive (HNO$_3$), the current Fe(III) catalytic system shows significant increase in the total yield, whereas the V(V) systems are effective under additive-free conditions [27,42]. The total yield under US conditions is also higher than that shown by a hydrazone Cu(II) catalytic system under MW conditions [27]. The selectivity towards CyOH of the current catalytic system is moderate in comparison with the hydrazone V(V) system [35,42].

From radical trapping experiments (addition of 2,2,6,6-tetramethylpiperidin-1-oxyl free radical to the reaction medium), in the presence of **1**, **2**, or **3** (Table 2, entries 21, 24, and 27, respectively), an extensive product yield inhibition (over 80%), relative to the total yield obtained when the reactions were carried out without any additive (Table 2, entries 5, 22, and 25, respectively), was observed. This suggests that the cyclohexane oxidation catalysed by complexes **1-3** proceeds through a radical mechanism, as proposed in other cases, depicted in Equations (1)–(9) [27,35,36,54,62]. Firstly, the reaction proceeds through the iron-catalyzed decomposition of the oxidant, leading to the formation of t-BuOO$^\bullet$ and t-BuO$^\bullet$ radicals upon the reduction of Fe(III) and oxidation of Fe(II) species, according to reactions (1) and (2), respectively. Then, cyclohexyl radical (Cy$^\bullet$) formation takes place due to H-abstraction from CyH by t-BuO$^\bullet$ (reaction (3)). Upon the reaction with dioxygen, Cy$^\bullet$ forms CyOO$^\bullet$ (reaction (4)), and then upon H-abstraction from TBHP by CyOO$^\bullet$, CyOOH is formed (reaction (5)). In the reactions (6) and (7), the Fe-assisted decomposition of CyOOH produces CyO$^\bullet$ and CyOO$^\bullet$, which leads to the formation of cyclohexanol (CyOH) and cyclohexanone (Cy=O) (the desired products), according to reactions (8) and (9).

$$[Fe^{III}] + t\text{-BuOOH} \rightarrow t\text{-BuOO}^\bullet + H^+ + [Fe^{II}] \tag{1}$$

$$[Fe^{II}] + t\text{-BuOOH} \rightarrow t\text{-BuO}^\bullet + [Fe^{III}] + HO^- \tag{2}$$

$$t\text{-BuO}^\bullet + \text{CyH} \rightarrow t\text{-BuOH} + \text{Cy}^\bullet \qquad (3)$$

$$\text{Cy}^\bullet + \text{O}_2 \rightarrow \text{CyOO}^\bullet \qquad (4)$$

$$\text{CyOO}^\bullet + t\text{-BuOOH} \rightarrow \text{CyOOH} + t\text{-BuOO}^\bullet \qquad (5)$$

$$\text{CyOOH} + [\text{Fe}^{II}] \rightarrow \text{CyO}^\bullet + \text{HO}^- + [\text{Fe}^{III}] \qquad (6)$$

$$\text{CyOOH} + [\text{Fe}^{III}] \rightarrow \text{CyOO}^\bullet + \text{H}^+ + [\text{Fe}^{II}] \qquad (7)$$

$$\text{CyO}^\bullet + \text{CyH} \rightarrow \text{CyOH} + \text{Cy}^\bullet \qquad (8)$$

$$2\text{CyOO}^\bullet \rightarrow \text{CyOH} + \text{Cy=O} + \text{O}_2 \qquad (9)$$

3. Experimental

3.1. General Materials and Procedures

The syntheses were performed in air. Reagents and solvents were obtained from commercial sources and used as received, without further purification or drying. Fe(NO$_3$)$_3$·9H$_2$O was used as the source of metal ion for the synthesis of **1**. Anhydrous FeCl$_2$ was used for the syntheses of **2** and **3**.

C, H, and N elemental analyses were carried out by the Microanalytical Service of the Instituto Superior Técnico. Infrared spectra (4000–400 cm^{-1}) were recorded on a Bruker Vertex 70 instrument (Bruker Corporation, Ettlingen, Germany) in KBr pellets; wavenumbers are given in cm^{-1}. Mass spectra were run in a Varian 500-MS LC Ion Trap Mass Spectrometer (Agilent technologies, Amstelveen, The Netherlands) equipped with an electrospray (ESI) ion source. The drying gas and flow rate were optimized (for electrospray ionization) according to the particular sample, with a 35 p.s.i. nebulizer pressure. Scanning was performed from m/z 100 to 1200 in methanol solution. The compounds were observed in the positive mode (capillary voltage = 80–105 V).

3.2. Synthesis of the Pro-Ligand H$_2$L

N'-acetylpyrazine-2-carbohydrazide (H$_2$L), the pro-ligand, (Scheme 1), was synthesized as described in the literature [45], by the acetylation of pyrazine-2-carbohydrazide.

Yield: 86.0%. Anal. calc. for (C$_7$H$_8$N$_4$O$_2$): C, 46.67; H, 4.48; N, 31.10; found: C, 46.62; H, 4.43; N, 31.06%. IR (KBr pellet, cm^{-1}): 3324 ν(NH), 3216 ν(NH), 1698 ν(C=O), 1670 ν(C=O). ^{1}H NMR (DMSO-d_6, δ): 9.18-8.86 (m, 3H, C$_4$H$_3$N$_2$), 8.74 (s, 2H, NH), 1.91 (s, 3H, CH$_3$).

3.3. Syntheses of Fe(III) Complexes of N'-acetylpyrazine-2-carbohydrazide

3.3.1. [Fe(HL)(H$_2$O)$_2$(NO$_3$)]NO$_3$ (1)

In this study, 0.404 g (1.00 mmol) of Fe(NO$_3$)$_3$·9H$_2$O was added to a 30 mL methanol solution of H$_2$L (0.180 g, 1.00 mmol). Then, at room temperature, the reaction mixture was stirred for 30 min, in open air. The resultant reddish brown solution was filtered and the filtrate was kept in air. Single crystals of **1** were isolated after 3 days, washed twice with cold CHCl$_3$, and dried in open air.

Yield 66%. Anal. Calcd. for C$_7$H$_{11}$FeN$_6$O$_{10}$: C, 21.28; H, 2.81; N, 21.27. Found: C, 21.20; H, 2.72; N, 21.18. IR (KBr pellet, cm^{-1}): 3136 ν(NH), 1383 ν(NO$_3$)$^-$, 1678 ν(C=O), 1646 ν(C=O), 1032 ν(N–N). ESI-MS (+): m/z 333 [M-(NO$_3$)]$^+$ (100%).

3.3.2. [Fe(HL)Cl$_2$] (2) and [Fe(HL)Cl(μ-OMe)]$_2$ (3)

Both **2** and **3** were synthesized according to the literature [45].

[Fe(HL)Cl$_2$] (2)

Yield 70%. Anal. Calcd. for C$_7$H$_7$Cl$_2$FeN$_4$O$_2$: C, 27.48; H, 2.31; N, 18.31. Found: C, 27.42; H, 2.28; N, 18.25. IR (KBr pellet, cm^{-1}): 3134 ν(NH), 1702 ν(C=O), 1668 ν(C=O), 1037 ν(N–N). ESI-MS (+): m/z 306 [M+H]$^+$ (100%).

[Fe(HL)Cl(μ-OMe)]$_2$ (3)

Yield 65%. Anal. Calcd. for C$_{16}$H$_{20}$Cl$_2$Fe$_2$N$_8$O$_6$: C, 31.87; H, 3.34; N, 18.58. Found: C, 31.84; H, 3.33; N, 18.54. IR (KBr; cm^{-1}): 3138 ν(NH), 1664 ν(C=O), 1641 ν(C=O), 1038 ν(N–N). ESI-MS (+): m/z 604 [M+H]$^+$ (100%).

3.4. X-ray Measurements

A crystal of **1** was immersed in cryo-oil, mounted in a Nylon loop, and measured at 150 K. Intensity data were collected using a Bruker AXS-KAPPA APEX II diffractometer (Bruker AXS Inc., Madison, WI, USA) with graphite monochromated Mo-Kα (λ 0.71073) radiation. Omega scans of 0.5° per frame were used for data collection and a full sphere of data was obtained. Cell parameters were retrieved using Bruker SMART [63] software and the data were refined using Bruker SAINT [63] on all the observed reflections. SADABS was used for absorption corrections [64]. Direct methods were employed by using SHELXS97 [65] and refined with SHELXL2018/3 [66]. Calculations were conducted using WinGX version 2018.3 [67]. The H-atoms bonded to carbon and nitrogen were introduced in the model at geometrically calculated positions and refined using a riding model, with U$_{iso}$ defined as 1.2U$_{eq}$ of the parent carbon atoms for aromatic residues and for nitrogen, and 1.5U$_{eq}$ for the methyl group. The hydrogen atoms of water ligands were found in the difference Fourier map and allowed to refine with distance restrains. The non-coordinated water molecule was disordered over two positions and refined with the use of PART instruction, with the occupancy of O1W and O2W at a ratio of 74% and 26%, respectively. After this strategy, the remaining electron density was considered as being due to another water molecule (O3W), which was refined with an occupancy of 0.25, flanked by PART -1, and the structure finalized normally. The hydrogen atoms of O1W, O2W, and O3W could be neither located nor inserted in the calculated positions (the use of the CALC-OH routine in WinGX proved to be unsuccessful). Least square refinements were carried out with anisotropic thermal motion parameters for all of the non-hydrogen atoms.

3.5. Catalytic Studies

The microwave-assisted peroxidative oxidation of cyclohexane was carried out in G10 Pyrex tubes (10 mL capacity reaction tube with a 13 mm internal diameter) in a focused Anton Paar Monowave 300 reactor (Anton Paar GmbH, Graz, Austria) fitted with a rotational system and an IR temperature detector.

The test was performed under the following conditions: The desired amount of catalyst **1**, **2**, or **3** (2.5–20 μmol); cyclohexane (5 mmol); acetonitrile (3 mL) and oxidant (1, 2, or 4 equivalent vs. substrate, *t*-BuOOH, 70% aqueous solution) were added into an Pyrex tube, which was introduced in the MW reactor and stirred under irradiation (5—10 W) at 50 °C. After completion of the desired reaction time, the reaction mixture was cooled to room temperature, whereafter 90 μL of cycloheptanone (internal standard) and 10 mL of diethyl ether (for substrate and organic product extraction) were added. The reaction mixture was stirred and centrifuged. A sample was taken from the mixture and analyzed by gas chromatography (GC) after the addition of an excess of triphenylphosphine, in order to reduce cyclohexyl hydroperoxide to cyclohexanol, following a method developed by Shul'pin [54]. The composition of products was confirmed by Gas Chromatography-Mass Spectrometry (GC-MS).

For the ultrasound-assisted oxidation reactions, a reaction tube was immersed in an ATU ultrasonic thermoregulated bath (40 kHz, 600 W) equipped with an automatic temperature heating-cooling circulatory system, which kept the bath temperature at ca. 50 °C for all of the trials.

GC measurements were carried out using a FISONS Instruments GC 8000 series gas chromatograph (Agilent Technologies, Santa Clara, CA, USA) with an FID detector and a capillary column (DB-WAX, column length: 30 m; internal diameter: 0.32 mm), using helium as a carrier gas and the Jasco-Borwin v.1.50 software (Jasco, Tokyo, Japan). The samples were injected at 240 °C, whereas the initial temperature was maintained at 100 °C for 1 min, increased to 180 °C at the rate 10 °C/min, and then hold for 1 min.

The samples were analysed by GC-MS using a Perkin Elmer Clarus 600 C instrument (Shelton, CT, USA) (He as the carrier gas). The ionization voltage was 70 eV. Gas chromatography was guided in the temperature-programming mode, using an SGE BPX5 column (30 m × 0.25 mm × 0.25 μm).

Retention times of all identified reaction products were compared with those of the commercially available samples. Reaction products' mass spectra were compared to fragmentation patterns obtained from the NIST spectral library stored in the computer software of the mass spectrometer.

4. Conclusions

Three Fe(III) compounds (**1-3**) derived from N'-acetylpyrazine-2-carbohydrazide were successfully applied for the peroxidative oxidation of cyclohexane under different energy stimuli (microwave irradiation, ultrasound, and conventional heating). In order to improve the catalytic performance, the effects of different reaction parameters were studied, namely the reaction time, catalyst amount, substrate:oxidant ratio, and presence of additives. Compounds **1-3** catalyzed the oxidation of cyclohexane via a radical mechanism, yielding cyclohexanol and cyclohexanone as the main products; the former with a high selectivity upon the reduction of the primary product cyclohexylperoxide (up to 95%). Complexes **1** and **3** exhibited similar catalytic activities, with 14–15% yields of cyclohexanol and cyclohexanone under 3 h of microwave irradiation at 50 °C, whereas complex **2** reached a total product yield of 10.4%.

Apart from being undertaken under microwaves, the peroxidative oxidation of cyclohexane was also performed using other types of energy inputs, i.e., conventional heating and ultrasound irradiation. The catalysts responded differently to the various energy stimuli, with the best performance of **1** being observed under microwave radiation, whereas complex **2** showed the maximum activity under conventional heating and compound **3** under ultrasounds. However, in all cases, the highest selectivity to cyclohexanol was verified when the reaction was assisted by microwave radiation.

In the case of **1**, a significant increase of the total yield was observed from 15 to 38% in the presence of an additive (HNO_3), although such an influence was not found for **2** and **3**. Therefore, the combined use of nitric acid and the mononuclear cationic Fe(III) compound **1** plays a crucial role in accelerating the catalytic oxidation process.

This study can help to foster the fruitful use of the environmentally acceptable oxidant aqueous TBHP and the application of microwave heating or ultrasounds to promote catalysis, which have significant environmental implications.

Supplementary Materials: CCDC number 2019464 contains the supplementary crystallographic data for **1**. This can be obtained free of charge via http://www.ccdc.cam.ac.uk/conts/retrieving.html, or from the Cambridge Crystallographic Data Center, 12 Union Road, Cambridge CB2 1EZ, UK; fax: (+44)-1223-336-033; or e-mail: deposit@ccdc.cam.ac.uk.

Author Contributions: Conceptualization, M.S. and E.C.B.A.A.; Methodology, M.S. and E.C.B.A.A.; Software, M.S., M.d.F.C.G.d.S. and T.R.B.; Validation, M.S., M.d.F.C.G.d.S. and E.C.B.A.A.; Formal Analysis, M.S. and T.R.B.; Investigation, M.S. and T.R.B; Resources, M.S. and A.J.L.P.; Data Curation, M.S., M.d.F.C.G.d.S. and T.R.B.; Writing-Original Draft Preparation, M.S., T.R.B. and E.C.B.A.A.; Writing-Review & Editing, M.d.F.C.G.d.S. and A.J.L.P.; Visualization, M.S.; Supervision, M.S. and E.C.B.A.A.; Project Administration, A.J.L.P.; Funding Acquisition, A.J.L.P and E.C.B.A.A. All authors have read and agreed to the published version of the manuscript.

Funding: This work has been supported by the Fundação para a Ciência e Tecnologia (FCT) 2020-2023 multiannual funding to Centro de Química Estrutural (project UIDB/00100/2020).

Acknowledgments: The authors are grateful to the Fundação para a Ciência e Tecnologia (FCT) project PTDC/QUI-QIN/29778/2017 for financial support. M.S. acknowledges the FCT and IST for a working contract "DL/57/2017" (Contract no. IST-ID/102/2018).

Conflicts of Interest: The authors declare no conflict of interest.

References

1. Pombeiro, A.J.L.; Guedes da Silva, M.F.C. (Eds.) *Alkane Functionalization*; Wiley: Hoboken, NJ, USA, 2019.

2. Pombeiro, A.J.L. Alkane Functionalization: Introduction and overview. In *Alkane Functionalization*; Pombeiro, A.J.L., Guedes da Silva, M.F.C., Eds.; Wiley: Hoboken, NJ, USA, 2019; Chapter 1; pp. 1–15.
3. Shilov, A.E.; Shul'pin, G.B. Activation of C-H bonds by metal complexes. *Chem. Rev.* **1997**, *97*, 2879–2932. [CrossRef] [PubMed]
4. Shul'pin, G.B. Selectivity enhancement in functionalization of C-H bonds: A review. *Org. Biomol. Chem.* **2010**, *8*, 4217–4228. [CrossRef] [PubMed]
5. Sutradhar, M.; Martins, L.M.D.R.S.; Guedes da Silva, M.F.C.; Pombeiro, A.J.L. Vanadium complexes: Recent progress in oxidation catalysis. *Coord. Chem. Rev.* **2015**, *301*, 200–239. [CrossRef]
6. Shul'pin, G.B. *Transition Metals for Organic Synthesis*, 2nd ed.; Beller, M., Bolm, C., Eds.; Wiley-VCH: New York, NY, USA, 2004; Volume 2, Chapter 2; pp. 215–242.
7. Shilov, A.E.; Shul'pin, G.B. *Activation and Catalytic Reactions of Saturated Hydrocarbons in the Presence of Metal Complexes*; Kluwer Academic Publishers: Dordrecht, The Netherlands, 2000.
8. Crabtree, R.H. Alkane C-H activation and functionalization with homogeneous transition metal catalysts: A century of progress—A new millennium in prospect. *J. Chem. Soc. Dalton Trans.* **2001**, *17*, 2437–2450. [CrossRef]
9. Derouane, E.G.; Haber, J.; Lemos, F.; Ribeiro, F.R.; Guinet, M. (Eds.) *Catalytic Activation and Functionalization of Light Alkanes*; NATO ASI Series; Kluwer Academic Publisher: Dordrecht, The Netherlands, 1998; Volume 44.
10. Retcher, B.; Sánchez Costa, J.; Tang, J.; Hage, R.; Gamez, P.; Reedijk, J. Unexpected high oxidation of cyclohexane by Fe salts and dihydrogen peroxide in acetonitrile. *J. Mol. Catal. A Chem.* **2008**, *286*, 1–5. [CrossRef]
11. Antony, R.; Manickam, T.S.; Kollu, P.; Chandrasekar, P.V.; Karuppasamy, K.; Balakumar, S. Highly dispersed Cu(II), Co(II) and Ni(II) catalysts covalently immobilized on imine-modified silica for cyclohexane oxidation with hydrogen peroxide. *RSC Adv.* **2014**, *4*, 24820–24830. [CrossRef]
12. Rahman, A.; Mupa, M.; Mahamadi, C. A mini review on new emerging trends for the synthesis of adipic acid from metal-nano heterogeneous catalysts. *Catal. Lett.* **2016**, *146*, 788–799. [CrossRef]
13. Guo, X.; Xu, M.; She, M.; Zhu, Y.; Shi, T.; Chen, Z.; Peng, L.; Guo, X.; Lin, M.; Ding, W. Morphology-reserved synthesis of discrete nanosheets of CuO@SAPO-34 and pore mouth catalysis for one-pot oxidation of cyclohexane. *Angew. Chem. Int. Ed. Engl.* **2020**, *59*, 2606–2611. [CrossRef]
14. Schuchardt, U.; Cardoso, D.; Sercheli, R.; Pereira, R.; Cruz, R.S.; Guerreiro, M.C.; Pires, E.L. Cyclohexane oxidation continues to be a challenge. *Appl. Catal. A Gen.* **2001**, *211*, 1–17. [CrossRef]
15. Pokutsa, A.; Le Bras, J. Muzart, Glyoxal-promoted homogeneous catalytic oxygenation of cyclohexane with hydrogen peroxide in the presence of V and Co compounds. *J. Russ. Chem. Bull. Int. Ed.* **2005**, *54*, 312–315. [CrossRef]
16. Pombeiro, A.J.L. (Ed.) *Advances in Organometallic Chemistry and Catalysis, The Silver/Gold Jubilee ICOMC Celebratory Book*; J.Wiley & Sons: New York, NY, USA, 2014.
17. Shul'pin, G.B. Hydrocarbon oxygenations with peroxides catalyzed by metal compounds. *Mini Rev. Org. Chem.* **2009**, *6*, 95–104. [CrossRef]
18. Li, J.J. (Ed.) *C-H Bond Activation in Organic Synthesis*; CRC Press: Boca Raton, FL, USA, 2015.
19. Pérez, P.J. (Ed.) *Alkane C-H Activation by Single-Site Metal Catalysis*; Springer: Berlin, Germany, 2012.
20. Bäckvall, J.-E. *Modern Oxidation Methods*; Wiley: Hoboken, NJ, USA, 2011.
21. White, M.C. Chemistry. Adding aliphatic C-H bond oxidations to synthesis. *Science* **2012**, *335*, 807–809. [CrossRef] [PubMed]
22. Newhouse, T.; Baran, P.S. If C-H bonds could talk: Selective C-H bond oxidation. *Angew. Chem. Int. Ed. Engl.* **2011**, *123*, 3422–3435. [CrossRef]
23. Olah, G.A.; Molnar, A.; Surya Prakash, G.K. *Hydrocarbon Chemistry*, 3rd ed.; Wiley: Hoboken, NJ, USA, 2017.
24. Sutradhar, M.; Martins, L.M.D.R.S.; Guedes da Silva, M.F.C.; Liu, C.-M.; Pombeiro, A.J.L. Trinuclear Cu(II) structural isomers: Coordination, magnetism, electrochemistry and catalytic activity toward oxidation of alkanes. *Eur. J. Inorg. Chem.* **2015**, *2015*, 3959–3969. [CrossRef]
25. Sutradhar, M.; Roy Barman, T.; Alegria, E.C.B.A.; Guedes da Silva, M.F.C.; Liu, C.-M.; Kou, H.-Z.; Pombeiro, A.J.L. Cu(II) complexes of N-rich aroylhydrazone: Magnetism and catalytic activity towards microwave-assisted oxidation of xylenes. *Dalton Trans.* **2019**, *48*, 12839–12849. [CrossRef] [PubMed]

26. Sutradhar, M.; Alegria, E.C.B.A.; Guedes da Silva, M.F.C.; Martins, L.M.D.R.S.; Pombeiro, A.J.L. Aroylhydrazone Cu(II) complexes in keto form: Structural characterization and catalytic activity towards cyclohexane oxidation. *Molecules* **2016**, *21*, 425. [CrossRef]
27. Sutradhar, M.; Alegria, E.C.B.A.; Guedes da Silva, M.F.C.; Liu, C.-M.; Pombeiro, A.J.L. Peroxidative oxidation of alkanes and alcohols under mild conditions by di- and tetranuclear copper(II) complexes of bis(2-hydroxybenzylidene)isophthalohydrazide. *Molecules* **2018**, *23*, 2699. [CrossRef]
28. Bonchio, M.; Carraro, M.; Scorrano, G.; Kortz, U. Microwave-assisted fast cyclohexane oxygenation catalyzed by iron-substituted polyoxotungstates. *Adv. Synth. Catal.* **2005**, *347*, 1909–1912. [CrossRef]
29. Carvalho, N.M.; Alvarez, H.M.; Horn, A., Jr.; Antunes, O.A. Influence of microwave irradiation in the cyclohexane oxidation catalyzed by Fe(III) complexes. *Catal. Today* **2008**, *133*, 689–694. [CrossRef]
30. Fernandes, R.; Lasri, J.; Guedes da Silva, M.F.C.; da Silva, J.A.L.; Pombeiro, A.J.L. Bis- and tris-pyridyl amino and imino thioether Cu and Fe complexes. Thermal and microwave-assisted peroxidative oxidations of 1-phenylethanol and cyclohexane in the presence of various N-based additives. *J. Mol. Catal. A Chem.* **2011**, *351*, 100–111. [CrossRef]
31. Ribeiro, A.P.C.; Martins, L.M.D.R.S.; Kuznetsov, M.L.; Pombeiro, A.J.L. Tuning cyclohexane oxidation: Combination of microwave irradiation and ionic liquid with the C-scorpionate [FeCl$_2$(Tpm)] catalyst. *Organometallics* **2017**, *36*, 192–198. [CrossRef]
32. Ribeiro, A.P.C.; Martins, L.M.D.R.S.; Carabineiro, S.A.C.; Buijnsters, J.G.; Figueiredo, J.L.; Pombeiro, A.J.L. Heterogenised C-scorpionate iron(II) complex on nanostructured carbon materials as catalysts for microwave-assisted oxidation reactions. *ChemCatChem* **2018**, *10*, 1821–1828. [CrossRef]
33. Ribeiro, A.P.C.; Matias, I.A.S.; Alegria, E.C.B.A.; Ferraria, A.M.; Botelho do Rego, A.M.; Pombeiro, A.J.L.; Martins, L.M.D.R.S. New trendy magnetic C-scorpionate iron catalyst and its performance towards cyclohexane oxidation. *Catalysts* **2018**, *8*, 69. [CrossRef]
34. Sutradhar, M.; Kirillova, M.V.; Guedes da Silva, M.F.C.; Martins, L.M.D.R.S.; Pombeiro, A.J.L. A hexanuclear mixed-valence oxovanadium(IV,V) complex as a highly efficient alkane oxidation catalyst. *Inorg. Chem.* **2012**, *51*, 11229–11231. [CrossRef] [PubMed]
35. Sutradhar, M.; Martins, L.M.; Roy, B.T.; Kuznetsov, M.L.; Guedes da Silva, M.F.C.; Pombeiro, A.J.L. Vanadium complexes of different nuclearities in the catalytic oxidation of cyclohexane and cyclohexanol—An experimental and theoretical investigation. *New. J. Chem.* **2019**, *43*, 17557–17570. [CrossRef]
36. Sutradhar, M.; Martins, L.M.D.R.S.; Carabineiro, S.A.C.; Guedes da Silva, M.F.C.; Buijnsters, J.G.; Figueiredo, J.L.; Pombeiro, A.J.L. Oxidovanadium(V) complexes anchored on carbon materials as catalysts for the oxidation of 1-phenylethanol. *ChemCatChem* **2016**, *8*, 2254–2266. [CrossRef]
37. Sutradhar, M.; Martins, L.M.D.R.S.; Guedes da Silva, M.F.C.; Alegria, E.C.B.A.; Liu, C.-M.; Pombeiro, A.J.L. Mn(II,II) complexes: Magnetic properties and microwave assisted oxidation of alcohols. *Dalton Trans.* **2014**, *43*, 3966–3977. [CrossRef]
38. Sutradhar, M.; Roy Barman, T.; Pombeiro, A.J.L.; Martins, L.M.D.R.S. Catalytic activity of polynuclear vs. dinuclear aroylhydrazone Cu(II) complexes in microwave-assisted oxidation of neat aliphatic and aromatic hydrocarbons. *Molecules* **2019**, *24*, 47. [CrossRef]
39. Sutradhar, M.; Alegria, E.C.B.A.; Mahmudov, K.T.; Guedes da Silva, M.F.C.; Pombeiro, A.J.L. Iron(III) and cobalt(III) complexes with both tautomeric (keto and enol) forms of aroylhydrazone ligands: Catalysts for the microwave assisted oxidation of alcohols. *RSC Adv.* **2016**, *6*, 8079–8088. [CrossRef]
40. Zaltariov, M.-F.; Alexandru, M.; Cazacu, M.; Shova, S.; Novitchi, G.; Train, C.; Dobrov, A.; Kirillova, M.V.; Alegria, E.C.B.A.; Pombeiro, A.J.L.; et al. Tetranuclear copper(II) complexes with macrocyclic and open-chain disiloxane ligands as catalyst precursors for hydrocarboxylation and oxidation of alkanes and 1-phenylethanol. *Eur. J. Inorg. Chem.* **2014**, *29*, 4946–4956. [CrossRef]
41. Dobrov, A.; Darvasiová, D.; Zalibera, M.; Bučinský, L.; Puškárová, I.; Rapta, P.; Martins, L.M.D.R.S.; Pombeiro, A.J.L.; Arion, V.B. Nickel(II) complexes with redox noninnocent octaazamacrocycles as catalysts in oxidation reactions. *Inorg. Chem.* **2019**, *58*, 11133–11145. [CrossRef]
42. Dragancea, D.; Talmaci, N.; Shova, S.; Novitchi, G.; Darvasiová, D.; Rapta, P.; Breza, M.; Galanski, M.; Kožíšek, J.; Martins, N.M.R.; et al. Vanadium(V) complexes with substituted 1,5-bis(2-hydroxybenzaldehyde)carbohydrazones and their use as catalyst precursors in oxidation of cyclohexane. *Inorg. Chem.* **2016**, *55*, 9187–9203. [CrossRef] [PubMed]

43. Arion, V.B.; Platzer, S.; Rapta, P.; Machata, P.; Breza, M.; Vegh, D.; Dunsch, L.; Telser, J.; Shova, S.; Mac Leod, T.; et al. Marked stabilization of redox states and enhanced catalytic activity in galactose oxidase models based on transition metal S-methylisothiosemicarbazonates with—SR group in ortho-position to the phenolic oxygen. *Inorg. Chem.* **2013**, *52*, 7524–7540. [CrossRef] [PubMed]
44. Dobrov, A.; Fesenko, A.; Yankov, A.; Stepanenko, I.; Darvasiová, D.; Breza, M.; Rapta, P.; Martins, L.M.D.R.S.; Pombeiro, A.J.L.; Shutalev, A.; et al. Nickel(II), Copper(II) and Palladium(II) complexes with Bis-Semicarbazide hexaazamacrocycles: Redox-noninnocent behavior and catalytic activity in oxidation and C-C coupling reactions. *Inorg. Chem.* **2020**, *59*, 10650–10664. [CrossRef] [PubMed]
45. Roy Barman, T.; Sutradhar, M.; Alegria, E.C.B.A.; Guedes da Silva, M.F.C.; Kuznetsov, M.L.; Pombeiro, A.J.L. Efficient Solvent-Free Friedel-Crafts Benzoylation and Acylation of *m*-Xylene Catalyzed by *N*-acetylpyrazine-2-carb hydrazide-Fe(III)-chloro Complexes. *Chem. Select* **2018**, *3*, 8349–8355.
46. Dudley, G.B.; Richert, R.; Stiegman, A.E. On the existence of and mechanism for microwave-specific reaction rate enhancement. *Chem. Sci.* **2015**, *6*, 2144–2152. [CrossRef]
47. Varma, R.S. Journey on greener pathways: From the use of alternate energy inputs and benign reaction media to sustainable applications of nano-catalysts in synthesis and environmental remediation. *Green Chem.* **2014**, *16*, 2027–2041. [CrossRef]
48. Ribeiro, A.P.C.; Alegria, E.C.B.A.; Palavra, A.; Pombeiro, A.J.L. Alkane functionalization under unconventional conditions: In ionic liquid, in supercritical CO_2 and microwave assisted. In *Alkane Functionalization*; Pombeiro, A.J.L., Guedes da Silva, M.F.C., Eds.; Wiley: Hoboken, NJ, USA, 2019; Chapter 24; pp. 523–537.
49. Ribeiro, A.P.C.; Alegria, E.C.B.A.; Kopylovich, M.N.; Ferraria, A.M.; Botelho do Rego, A.M.; Pombeiro, A.J.L. On the comparison of microwave and mechanochemical energy inputs in catalytic oxidation of cyclohexane. *Dalton Trans.* **2018**, *47*, 8193–8198. [CrossRef]
50. Perkas, N.; Wang, Y.; Koltypin, Y.; Gedanken, A.; Chandrasekaran, S. Mesoporous iron–titania catalyst for cyclohexane oxidation. *Chem. Commun.* **2001**, 988–989. [CrossRef]
51. Sutradhar, M.; Roy Barman, T.; Pombeiro, A.J.L.; Martins, L.M.D.R.S. Cu(II) and Fe(III) complexes derived from N-acetylpyrazine-2-carbohydrazide as efficient catalysts towards solvent-free microwave assisted oxidation of alcohols. *Catalysts* **2019**, *9*, 1053. [CrossRef]
52. Sutradhar, M.; Guedes da Silva, M.F.C.; Pombeiro, A.J.L. Synthesis and chemical reactivity of an Fe(III) metallacrown-6 towards N-donor Lewis bases. *Inorg. Chem. Commun.* **2013**, *30*, 42–45. [CrossRef]
53. Sutradhar, M.; Guedes da Silva, M.F.C.; Nesterov, D.S.; Jezierska, J.; Pombeiro, A.J.L. 1D coordination polymer with octahedral and square-planar nickel(II) centers. *Inorg. Chem. Commun.* **2013**, *29*, 82–84. [CrossRef]
54. Shul'pin, G.B.; Nizova, G.V. Formation of alkyl peroxides in oxidation of alkanes by H_2O_2 catalyzed by transition metal complexes React. *Kinet. Catal. Lett.* **1992**, *48*, 333–338. [CrossRef]
55. Pokutsa, A.; Pawel Bloniarz, P.; Fliunt, O.; Kubaj, Y.; Zaborovskyia, A.; Paczeŝniakc, T. Sustainable oxidation of cyclohexane catalyzed by a VO(acac)$_2$-oxalic acid tandem: The electrochemical motive of the process efficiency. *RSC Adv.* **2020**, *10*, 10959–10971. [CrossRef]
56. Shul'pin, G.S.; Mishra, L.S.; Shul'pina, T.V.; Strelkova, A.J.L. Pombeiro. Oxidation of hydrocarbons with hydrogen peroxide catalysed by maltolato vanadium complexes covalently bonded to silica gel. *Catal. Commun.* **2007**, *8*, 1516–1520. [CrossRef]
57. Sutradhar, M.; Alegria, E.C.B.A.; Barman, T.R.; Scorcelletti, F.; Guedes da Silva, M.F.C.; Pombeiro, A.J.L. Microwave-assisted peroxidative oxidation of toluene and 1-phenylethanol with monomeric keto and polymeric enol aroylhydrazone Cu(II) complexes. *Mol. Catal.* **2017**, *439*, 224–232. [CrossRef]
58. Días-Ortiz, Á.; Prieto, P.; de la Hoz, A. A critical overview on the effect of microwave irradiation in organic synthesis. *Chem. Rec.* **2019**, *19*, 85–97. [CrossRef]
59. Herrero, M.A.; Kremsner, J.M.; Kappe, C.O. Nonthermal microwave effects revisited: On the importance of internal temperature monitoring and agitation in microwave chemistry. *J. Organomet. Chem.* **2007**, *73*, 36–47. [CrossRef]
60. Obermayer, D.; Kappe, C.O. On the importance of simultaneous infrared/fiber-optic temperature monitoring in the microwave-assisted synthesis of ionic liquids. *Org. Biomol. Chem.* **2010**, *8*, 114–121. [CrossRef]
61. Piermattei, A.; Karthikeyan, S.; Sijbesma, R.P. Activating catalysts with mechanical force. *Nat. Chem.* **2009**, *1*, 133–137. [CrossRef]
62. Shul'pin, G.B. Metal-catalysed hydrocarbon oxidations. *C. R. Chim.* **2003**, *6*, 163–178. [CrossRef]

63. Bruker AXS Inc. *Bruker, APEX2*; Bruker AXS Inc.: Madison, Wisconsin, USA, 2012.
64. Sheldrick, G.M. SADABS. In *Program for Empirical Absorption Correction*; University of Göttingen: Göttingen, Germany, 2000.
65. Sheldrick, G.M. SHELX97. In *Programs for Crystal Structure Analysis (Release 97-2)*; University of Göttingen: Göttingen, Germany, 1997.
66. Sheldrick, G.M. Crystal structure refinement with SHELXL. *Acta Cryst.* **2015**, *C71*, 3–8. [CrossRef] [PubMed]
67. Farrugia, L.J. wingx and ortep for windows: An update. *J. Appl. Cryst.* **2012**, *45*, 849–854. [CrossRef]

© 2020 by the authors. Licensee MDPI, Basel, Switzerland. This article is an open access article distributed under the terms and conditions of the Creative Commons Attribution (CC BY) license (http://creativecommons.org/licenses/by/4.0/).

Communication

Pd Nanoparticles and Mixture of $CO_2/CO/O_2$ Applied in the Carbonylation of Aniline

Dominik Madej, Adrian Konopko, Piotr Piotrowski and Agnieszka Krogul-Sobczak *

Faculty of Chemistry, University of Warsaw, Pasteura 1, 02-093 Warsaw, Poland; dmadej@connect.ust.hk (D.M.); akonopko@chem.uw.edu.pl (A.K.); ppiotrowski@chem.uw.edu.pl (P.P.)
* Correspondence: akrogul@chem.uw.edu.pl; Tel.: +48-22-552-6289; Fax: +48-22-822-5996

Received: 19 June 2020; Accepted: 28 July 2020; Published: 4 August 2020

Abstract: CO_2 is a compound of high stability which proves useful in some organic syntheses as a solvent or component decreasing explosivity of gases. It is also a good carbonylating agent for aliphatic amines although not for aromatic ones, the latter being carbonylated with phosgene or, as in our previous works, with CO/O_2 in the presence of Pd(II) complexes. In this work we have used the mixture of CO/O_2 and CO_2 for carbonylation of aniline to N,N'-diphenylurea. After optimization of the reaction conditions (56% of CO_2 in CO_2/CO mixture) we studied the activity of three kinds of pre-catalysts: (a) Pd(II) complexes, (b) Pd_{black}, and (c) palladium nanoparticles (PdNPs) in the presence of derivatives of pyridine (X_nPy). The highest conversion of aniline (with selectivity towards N,N-diphenylurea ca. 90%) was observed for PdNPs. The results show that catalytic cycle involves Pd(0) stabilized by pyridine ligand as active species. Basing on this observation, we put the hypothesis that application of PdNPs instead of Pd(II) complex can efficiently reduce the reaction time.

Keywords: Palladium; nanoparticles; carbonylation; aniline; carbon monoxide; carbon dioxide

---

1. Introduction

Environment-friendly organic synthesis is the main challenge of modern science in order to meet the objectives of sustainable development. Among many types of functionalization, insertion of a single carbonyl group into an organic molecule plays a special role due to large-scale demand for carbonyl compounds [1,2]. Excellent examples of products (both commodities and specialty chemicals) with carbonyl groups are urea derivatives, widely used in modern chemical industry— they find application in the production of pesticides (herbicides, fungicides), resin precursors, and fiber dyes [3,4], as well as antiviral and anticancer agents and other pharmaceuticals [4–8]. Numerous derivatives of ureas—including isocyanates and carbamates—are employed in syntheses of adhesives, varnishes, rubbers, paints, and polyurethane foams [9]. Unfortunately, dominating technologies of production of diphenylureas from aromatic amines are based on the phosgene method [9,10], and the real challenge in this sector of industry is to replace them by phosgene-free methods. The most common approaches involve less environmentally harmful carbonylating agents such as CO or alkyl carbonates [1,3,5,11–13], including carbonylation performed in beneficial nonconventional solvents such as ionic liquids [12,13]. The main limitation of CO-based methods is the high pressure applied.

Over the past few years, our studies have been focused on the carbonylation of aromatic nitrocompounds and amines by CO in the presence of the $PdCl_2(X_nPy)_2/Fe/I_2/X_nPy$ catalytic system, where Py = pyridine, X = Cl or CH_3, n = 0–2 [14–17]. We have successfully optimized the reaction conditions and proposed detailed mechanisms for carbonylation of aniline (AN) to N,N'-diphenylurea (DPU, equation 1), or to ethyl N-phenylcarbamate (EPC, Equation (2)), by CO/O_2.

$$2 \; C_6H_5NH_2 + CO + 1/2\,O_2 \xrightarrow{cat.} C_6H_5NH-CO-NHC_6H_5 + H_2O \qquad (1)$$

$$C_6H_5NH_2 + CO + ROH + 1/2\,O_2 \xrightarrow{cat.} C_6H_5NHCOOR + H_2O \qquad (2)$$

Processes presented by Equations (1) and (2) involve mixture of CO (a common source of carbonyl group) and O_2 (oxidizing agent), which is potentially explosive in a wide range of concentrations: 16.7–93.5% (under atmospheric pressure and at 18 °C) [18,19]. Therefore, strategies of eliminating or replacing hazardous substrates with safer and less expensive compounds are actively researched [20], and introduction of carbon dioxide to gaseous components is one of such approaches following the promising trends of green chemistry. CO_2 is already used as a reagent in relatively few industrial processes such as production of urea, salicylic acid, and some carbonates. Despite limited number of applications of CO_2 in organic synthesis caused mainly by high kinetic inertness and thermodynamic stability of CO_2 [21–24], every year new approaches involving CO_2 have been investigated [25–28]. For many years, carbonylation of aromatic amines by CO_2 was poorly represented in the literature, in contrast to carbonylation of ammonia and aliphatic amines by CO_2 [29–34]. Perhaps, lower nucleophilicity of nitrogen atom in the aromatic ring of aromatic amines decreases their reactivity towards CO_2 [5]. However, in recent years significant progress has been made toward the synthesis of isocyanates, ureas, carbamates, and other compounds using CO_2 and aromatic amines [35–41]. Despite promising results, many of these methods suffer from some limitations such as long reaction times required, harsh reaction conditions, low yields, and other difficulties in application of CO_2 as a carbonylating agent [42,43]. Therefore, some reports are focused on the use of CO_2 as an additive for CO, or as a reaction medium (liquid CO_2) [5,27,44]. Gabriele et al. [44] observed that in carbonylation of amines performed in the presence of CO/air and PdI_2, the addition of CO_2 significantly increased yield of the reaction for aliphatic amines, whereas less satisfactory results were obtained for carbonylation of aromatic amines. Surprisingly, good performance was observed for both aliphatic and aromatic amines when carbonylation was conducted in pure CO_2 as a non-polar aprotic solvent. Using an appropriate amount of CO_2 resulted in nearly three times higher catalytic activity of PdI_2 catalyst [44]. Based on the results obtained by Gabriele, although in many processes it is very difficult to replace CO by CO_2 as carbonylating agent, addition of CO_2 may increase the yield of the carbonylation of amines by CO. Moreover, carbon dioxide is a byproduct formed during industrial production of CO and its complete removal makes an additional complication in the production process. Furthermore, CO_2 exhibits a much stronger suppression effect on the explosion of flammable gases than nitrogen [45], and thus it decreases explosiveness of gases employed in the synthesis (CO and O_2) leading to enhanced safety of the process [46]. Last but not least, presence of CO_2 in a traditional liquid phase under mild pressures (tens of bar) results in generation of a gas-expanded liquid (GXL) phase. GXL retains the beneficial attributes of a conventional solvent (polarity, catalyst/reactant solubility) with some additional advantages: higher miscibility of permanent gases (O_2, CO, etc.) and enhanced transport rates compared to organic solvents at ambient conditions. The enhanced gas solubilities in GXLs may result in reaction rates greater than those achieved in neat organic solvent or supercritical carbon dioxide (sCO_2) [47]. In the case of our process, even if CO_2 cannot serve as a carbonylating agent, it is interesting to explore other potential benefits of replacing CO with CO_2, i.e., decreasing the amount of CO used and introducing CO_2 without any preconceived notion regarding its exact function.

Inspired by the promising properties of CO_2 and results reported by Gabriele et al. for carbonylation of aniline in the presence of CO_2, K_2PdI_4 as catalyst and without any additives [44], we decided to study the effect of CO_2 as one of the components of $CO/O_2/CO_2$ mixture on the carbonylation of aniline (model aromatic amine) in the presence of our original catalytic system $PdCl_2(X_nPy)_2/Fe/I_2/X_nPy$, at shorter reaction time and under lower total pressure. Also, our goal is to develop our previous catalytic system based on $PdCl_2(X_nPy)_2$ complexes into nanocatalysts, characterized by unique catalytic properties. Based on the results of our recent studies, we turned our attention to derivatives of pyridine as ligands that optimally stabilize palladium NPs [48–51] i.e., the access to the catalyst surface is not restricted, in contrast to bulky ligands [52–54]. Choosing 4-methylpyridine (model derivative of pyridine) as stabilizing ligand allows nanoparticles (NPs) to effectively interact with ligands and the reacting compound(s). In our previous works, we developed a reduction of aromatic nitrocompounds to aromatic amines in the presence of palladium nanoparticles stabilized by 4-methylpyridine (PdNPs/4MePy) [55,56]. Obtained results encourage us investigation of the catalytic activity of PdNPs/4-MePy in another process i.e., oxidative carbonylation of aniline in the presence of mixture $CO_2/CO/O_2$. For the first time, carbonylation of aniline is carried out in the presence of CO_2 and palladium nanoparticles stabilized by 4-methylpyridine.

2. Results and Discussion

Referring to our previous work [56], monodentate N-heterocyclic compounds are potentially a new family of stabilizing agents that could be a starting point for design new catalytically active nanoparticles with higher catalytic efficiency. In this work, the catalytic activity of PdNPs is compared with effectiveness of catalytic system based on Pd(II) complexes in the carbonylation of aniline (AN) to N,N'-diphenylurea (DPU) by $CO_2/CO/O_2$ mixture. Based on our previous results, we proposed the mechanism of AN carbonylation by CO/O_2 in the presence $PdCl_2(X_nPy)_2$ complexes (where: Py = pyridine, X = -Cl or -CH_3, n = 0–2), with Pd(II) reduced to Pd(0) in situ in the catalytic cycle, see Scheme 1. Partial precipitation of inactive Pd_{black} reported by Ragaini [57] is one of the proofs of the Pd^0 presence in the system, further supported by our isolation of palladium black precipitated during the reaction of $PdCl_2(PhNH_2)_2$ complex with carbon monoxide [15]. The next step is reoxidation of Pd(0) to Pd(II) and in this cycle both molecular oxygen and iodine are supposed to be potential oxidants responsible for recycling Pd(II) from Pd(0), step 1a–1b. Although oxidation of Pd(0) by oxygen is possible, it is very slow [58]. Alternatively, Pd(0) in 1a may be oxidized during the oxidative addition of I_2 to Pd(0), according to the equation: $Pd(0) + I_2 \rightarrow PdI_2$. Then, HI (instead of water) is released and this HI is oxidized by molecular oxygen: $4HI + O_2 \rightarrow 2I_2 + 2H_2O$ [1]. The intermediate 1b is able to coordinate aniline with subsequent insertion of CO to NH-Pd bond in 1c, creating a new carbon-nitrogen bond, intermediate 1d reductive elimination gives N,N'-diphenylurea, generating Pd(0) species [15]. According to the proposed mechanism presented in Scheme 1, Pd(0) stabilized by pyridine ligands plays a crucial role as an acceptor of oxidizing agent (O_2 or I_2). In order to verify the hypothesis on the participation of Pd(0) in the catalytic cycle, we planned experiments with Pd(0) nanoparticles (PdNPs), and the results (conversion, selectivity, and yield of carbonylation) were compared with the same parameters for process catalyzed by Pd(II) complexes. Prior to that, we searched for the optimal conditions to make both processes, catalyzed by PdNPs and Pd(II) complexes, comparable.

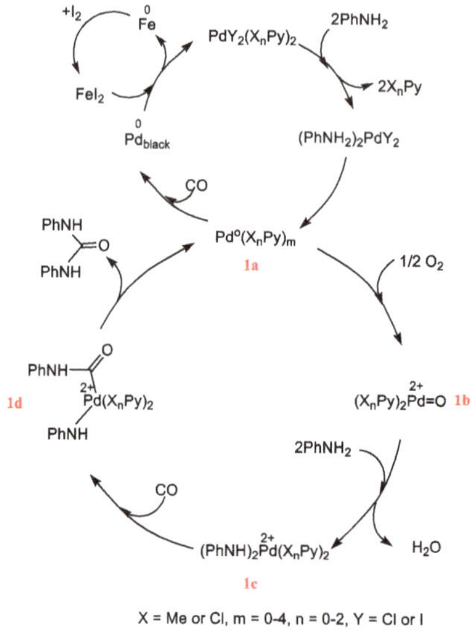

Scheme 1. Proposed path of the carbonylation of aniline by CO/O$_2$ catalyzed by PdCl$_2$(X$_n$Py)$_2$ complexes.

2.1. Optimization of Reaction Conditions of AN Carbonylation with CO$_2$/CO/O$_2$ Mixture

Conversion of aniline, selectivity towards DPU, and TOF for DPU are presented in Table 1, indicating a strong correlation between the rate of reaction and the amount of CO$_2$ used. Relatively high conversion and selectivity are observed when no CO$_2$ is loaded into the system (entry 5, Table 1). However, even a moderate addition of CO$_2$ is associated with the increasing rate of reaction, and its optimal amount in the gaseous mixture is ca. 50% (entry 4). High yield of DPU (96%) was also observed by Gabriele et al. for carbonylation of aniline performed in the presence of CO$_2$ and a different Pd-based catalytic system. Authors applied a very simple catalyst (K$_2$PdI$_4$), without any co-catalyst and additives; however, a long reaction time (24–72 h) was required [44]. Figure 1 shows that conversion of aniline and selectivity towards DPU decrease drastically when more than 50% of CO$_2$ is introduced. Eventually, no formation of DPU is observed at 100% fraction of CO$_2$ (entry 1, Table 1). Investigation of the impact of CO/CO$_2$ ratio was complemented by additional experiment in which carbonylation of aniline under conditions from entry 4 in Table 1 was conducted in the presence of Ar/CO/O$_2$ instead of CO$_2$/CO/O$_2$ mixture, and no difference in DPU yield between the two cases was observed. Obtained results suggest that replacing CO with certain amount of CO$_2$ allows for achieving higher yields of DPU, although, as indicated by the experiment with argon, CO$_2$ itself does not seem to act as a carbonylating agent. We propose two possible explanations for the beneficial effect of CO$_2$. First, it is likely that the observed optimal CO/CO$_2$ ratio is a result of the balance between enhancement of mass transfer by CO$_2$ (as commonly observed in GXLs) and minimal amount of CO necessary for efficient carbonylation of aniline. In the presence of 88% of CO$_2$ content, the amount of CO (12%) is lower than required according to stoichiometry, and therefore, unsurprisingly, is associated with lower conversion and TOF values. On the other hand, the presence of CO$_2$ might prevent (inhibit) possible side reactions, e.g., oxidation of CO to CO$_2$ (which to some extent may occur in the presence of palladium catalyst). Unless the necessary amount of CO is provided in the system, no DPU is produced and such results confirm that CO$_2$ is not a source of carbonyl group in this reaction. However, we performed further processes in the presence of CO$_2$ as an additional gas due to its economic benefits (CO$_2$ is a natural waste in CO production).

Table 1. Parameters obtained for the carbonylation [a] of aniline with $CO/CO_2/O_2$ mixture, catalyzed by Pd(II) complexes: conversion (C_{AN}), selectivity (S_{DPU}) and turnover frequency (TOF_{DPU}) of the catalyst, depending on the contents of CO_2 in CO_2/CO, iodine, and iron.

Entry	CO_2 in CO/CO_2 [b] (%)	I_2 (mmoL)	Fe (mmoL)	C_{AN} (%)	S_{DPU} [c] (%)	TOF_{DPU} [d]
1 [e]	100	0.12	2.68	12	0	0
2 [e]	88			37	57	202
3 [e]	65			83	83	666
4 [e]	56			84	87	704
5 [e]	0			68	93	608
6	56	0	2.68	10	0	0
7		0	0	10	0	0
8		0.01		46	96	434
9		0.04		71	96	657
10		0.12		73	95	669
11		0.39		69	94	617
12		0.04	0.25	75	96	694
13			0.5	78	98	733
14			1.2	74	96	685
15			2.68	72	95	660
16 [f]			2.68	72	96	667

[a] Reaction conditions unless stated otherwise: $PdCl_2(2,4Cl_2Py)_2$ = 0.056 mmoL, AN = 54 mmoL, ethanol = 20 mL, 3.8 MPa CO + CO_2, 0.6 MPa O_2, (100 °C, 60 min, Py = pyridine, AN = aniline, DPU = N,N-diphenylurea. [b] For CO/O_2: (0.6 MPa O_2 and 3.8 MPa CO) the molar ratio $CO:O_2$ = 6:1 ca. For $CO/O_2/CO_2$ (0.6 MPa O_2, 1.7 MPa CO, 2.1 MPa CO_2) the molar ratio $CO:O_2$ = 3:1 ca. [c] Selectivity toward DPU expressed as (mmoL DPU) × (mmoL converted AN)$^{-1}$ [%]. [d] TOF_{DPU} (turnover frequency for AN) = [mmoL of AN reacted selectively to DPU] × [mmoL of Pd(II) complex used]$^{-1}$ × h^{-1}, [e] $PdCl_2Py_2$ used instead of $PdCl_2(2,4Cl_2Py)_2$. [f] 0.6 mL of 2,4-Cl_2Py added.

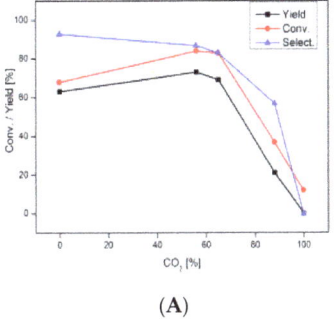

(A)

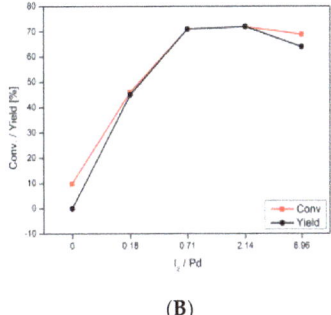

(B)

Figure 1. Effect of percentage of CO_2 in CO_2/CO mixture (**A**) and amount of iodine (**B**) on the conversion (Conv.) and yield and selectivity (Select.) of the catalyst. For the reaction conditions see footnotes in Table 1.

The substituent effect in the pyridine ring was investigated for the optimized content of CO_2 in CO_2/CO mixture (i.e., 56%). On the basis of obtained results, there is no consistent trend in the effect of derivatives of pyridine in $PdCl_2$ complexes on the rate of reaction (Figure S1 in Supplementary Material). The slightly higher yield of DPU (comparing to other Pd-based complexes) is noticed for $PdCl_2(2,4-Cl_2Py)_2$ complex, and most further studies in this area are performed in the presence of this complex.

The effect of iodine on the rate of carbonylation was also investigated. As shown in Table 1, when no iodine is used, the desired reaction does not proceed, regardless of the amount of iron present in the system (entries 6–7, Table 1). Increasing amount of iodine results in higher conversion, selectivity and TOF values, with the maximum at 0.04–0.12 mmoL of I_2 (entry 9 and 10, Table 1). This observation, in agreement with previous reports [1,59–62], indicates that iodine might play various crucial roles such as: (i) recovery of the catalytic system, perhaps by oxidation of iron powder to iron(II), (ii) generation of palladium complexes [Pd(CO)$_3$I]$^-$, considered to be the catalytically active species, (iii) reoxidation of Pd(0) to Pd(II). However, if a large excess of iodine is used, the yield of DPU decreases, and the TOF value decreases (entry 11, Table 1). This effect may be attributed to undesirable side reactions, such as formation of insoluble anilinium iodide, which limits the amount of free aniline in the system [60,63]. Moreover, excess of iodide ions may get coordinated to Pd(II), effectively competing with other reagents, which is commonly referred to as catalyst poisoning [10,60,64].

The effect of iron on the rate of reaction is minor (entries 12–16, Table 1). Although a small addition of iron to the mixture seems to increase the conversion of aniline and selectivity towards DPU (see entries 12 and 13, Table 1), it is not a significant change. As more iron is introduced to the system, conversion of aniline decreases slightly (entry 14, Table 1). These observations indicate that a certain amount of iron is beneficial and it slightly increases TOF values during DPU formation. In agreement with our report [15], Fe(0) is oxidized by I$_2$ to Fe(II) and the possible role of Fe(II) is to react with Pd$_{black}$ in order to return Pd$_{black}$ to the catalytic cycle as shown in Scheme 1. In the literature, we can find reports for [1] and against [65] the suggested role of iron. Lower conversion of aniline, when excess of iron is used, can be attributed to the reaction between iron and oxygen, which in turn decreases the amount of necessary oxidizing agent (O$_2$). Both reactions of metallic iron, with iodine and with O$_2$, occur easily [66]. Our previous research indicates that even traces of iron (from stainless steel reactor and stirring element) are kinetically significant.

The reaction rate seems to be strongly dependent on the temperature settings selected (see Table 2): at 80 °C the reaction proceeds only to some extent (entry 1). Optimal value for the synthesis of DPU is 100–120 °C (entry 2 and 3) with even higher temperature (140 °C) leading to formation of EPC (entry 4, Table 2, selectivity and TOF values for EPC are placed in parentheses). These results suggest that, after achieving the activation parameters suitable for the formation of DPU, further increase of temperature does not enhance the catalyst activity and may even have a slightly negative impact on the conversion of aniline, possibly due to occurrence of side reactions such as formation of N-ethylaniline, 2-methylquinoline, polyaniline, and EPC (entry 4, Table 2). Further studies were conducted at 100 °C because one of our aims was to operate at desirable energy-saving conditions, i.e., at the lowest temperature allowing formation of satisfactory amount of DPU. The temperature of 100 °C was also chosen for other practical reasons—it was the most appropriate temperature for comparative tests (high conversion and selectivity of Pd(II) complex were observed at this temperature during our previous studies reported in [15]). Although NPs might be more active at 120 °C than at 100 °C, the goal of this work was not to achieve the highest possible activity of catalyst but to study and compare activity of various types of pre-catalysts at the same temperature.

Table 2. Parameters obtained for the carbonylation of aniline by CO/CO$_2$/O$_2$ catalyzed by PdCl$_2$(2,4-Cl$_2$Py)$_2$: conversion (C$_{AN}$), selectivity (S$_{DPU}$) and turnover frequency (TOF$_{DPU}$), depending on the temperature [a].

Entry	T (°C)	C$_{AN}$ (%)	S$_{DPU}$ [b] (%)	TOF$_{DPU}$ [c]
1	80	17	98	161
2	100	71	98	671
3	120	86	95	791
4	140	80	71 (20) [d]	550 (154) [e]

[a] Reaction conditions unless stated otherwise: PdCl$_2$(2,4Cl$_2$Py)$_2$ = 0.056 mmoL, I$_2$ = 0.04 mmoL, AN = 54 mmoL, ethanol = 20 mL, 3.8 MPa CO + CO$_2$, 0.6 MPa O$_2$, 100 °C, 60 min, Py = pyridine, AN = aniline, DPU = N,N-diphenylurea. [b] Selectivity toward DPU expressed as (mmoL DPU) × (mmoL converted AN)$^{-1}$ [%]. [c] TOF$_{DPU}$ (turnover frequency for AN) = [mmoL of AN reacted selectively to DPU] × [mmoL of PdCl$_2$(2,4ClPy)$_2$]$^{-1}$ × h^{-1}. [d] Selectivity toward EPC (ethyl N-phenylcarbamate), [e] TOF$_{EPC}$ = (mmoL of EPC formed)(mmoL of PdCl$_2$(2,4-Cl$_2$Py) used)$^{-1}$h^{-1}.

2.2. Synthesis of PdNPs

PdNPs stabilized by 4MePy were synthesized in water, following the procedure [43,50]. Before NPs are used as catalysts they are dried because the presence of excess H_2O in the reaction mixture might diminish the yield of diphenylurea formed during the carbonylation of aniline by CO/O_2 (partial hydrolysis of urea may occur at high temperature) [1]. The TEM images presented in Figure 2 indicate agglomeration after drying (left panel versus middle panel), therefore, we also decided to synthesize NPs in ethanol. Right panel in Figure 2 demonstrates that palladium nanoparticles are not stable in ethanol and bigger aggregates are formed. A more appropriate name for these aggregates would be a palladium-based nanostructural material (PdNM) rather than nanoparticles.

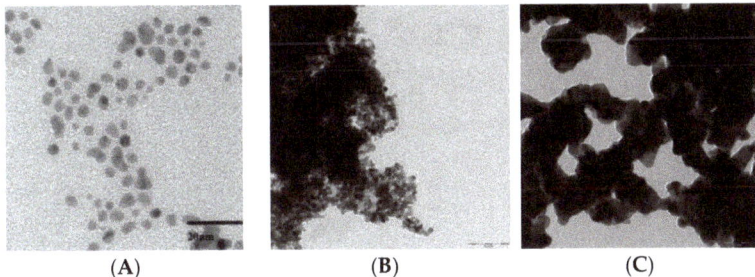

Figure 2. (**A**) TEM image of the PdNPs stabilized by 4-methylpyridine (PdNPs/4MePy) from raw aqueous solution [51]. Pd: $NaBH_4$ molar ratio = 1:2, concentration of $NaBH_4$ solution = 1%. For synthesis conditions see Experimental Section. (**B**) TEM image of the PdNPs stabilized by 4-methylpyridine (PdNPs/4MePy) dried and re-suspended in distilled-deionized water. (**C**) TEM image of the PdNM (palladium-based nanostructural material) stabilized by 4-methylpyridine (PdNM/4MePy) obtained in ethanol.

2.3. Catalytic Activity of NPs Compared with Other Pd Species

Conversion of aniline, selectivity towards DPU, and TOF for DPU formed in the presence of PdNPs/4MePy are presented in Table 3. Results obtained for PdNPs are compared with results obtained for commercially available Pd_{black} and two Pd(II) complexes: with 4MePy (the ligand that forms the most stable PdNPs) and with 2,4-Cl_2Py (the ligand forming the most catalytically active complex of Pd(II)). TOF values of PdNPs, Pd_{black}, and Pd(II) complex, measured after 60 min demonstrate that all studied substances are effective pre-catalysts for carbonylation of aniline during the standard time of reaction. The highest yield observed for PdNPs indicates that PdNPs are either catalytically active species or the most efficient source of other catalytically active species e.g., $[Pd(CO)_3I]^-$ [10]. Perhaps, when Pd(II) complex is applied as a pre-catalyst, it takes longer time to generate in situ catalytically active Pd(0) species from Pd(II), which might explain TOF for Pd(II) complexes being lower than for PdNPs. Our hypothesis, that active species responsible for catalytic activity are easily formed from PdNPs, is confirmed by results obtained for initial stages of the process (i.e., reaction carried out within first 15 min)-significant difference between TOF of PdNPs and TOF of Pd(II) complex is noticed. It is possible that, when carbonylation is carried out in the presence of PdNPs, catalytically active Pd(0) species are immediately present in the reaction from the beginning of the process. Differences in TOF values noticed for Pd(0) in the form of Pd_{black} (entry 4) and Pd(0) in the form of PdNM (entry 3) indicate that higher catalytic activity of PdNM might have an origin in the nanostructure of investigated material (image in the right panel of Figure 2 displays the PdNM formed from PdNPs with diameter ca. 10 nm). The lowest TOF value observed for Pd_{black} suggests that catalytically active species are not in the form of a heterogeneous bulk metal, but rather homogeneous complexes (with Pd^0), as suggested in the literature [60]. Such a hypothesis that homogeneous complexes (with Pd^0) are the real catalysts can explain why TOF for Pd(II) is higher than for Pd(0) –formation of homogeneous Pd(0) species from bulky Pd$_{black}$ is more difficult than from Pd(II).

Table 3. Parameters obtained for the carbonylation of aniline by CO/CO$_2$/O$_2$: conversion (C_{AN}), selectivity (S_{DPU}) and turnover frequency (TOF_{DPU}), depending on the catalyst [a].

Entry	Catalyst	C_{AN} (%)	S_{DPU} [b] (%)	TOF_{DPU} [c]
1	PdNPs/4MePy	92	88	781
2 [d]	PdNPs/4MePy	27	89	926
3	PdNM/4MePy [e]	80	84	646
4	Pd$_{black}$	60	89	515
5	Pd$_{black}$ [f]	72	89	617
6	PdCl$_2$(4MePy)$_2$	83	87	656
7 [d]	PdCl$_2$(2,4Cl$_2$Py)$_2$	16	88	540

Reaction conditions unless stated otherwise: Pd-based catalyst/Fe/I$_2$ = 0.056/0.5/0.12 mmoL, AN = 54 mmoL, 1.5 MPa CO, 1.9 MPa CO$_2$, 0.6 MPa O$_2$, 20 mL EtOH, 60 min. Py = pyridine, AN = aniline, DPU = N,N-diphenylurea. [b] Selectivity toward DPU expressed as (mmoL DPU) × (mmoL converted AN)$^{-1}$ [%]. [c] TOF$_{DPU}$ (turnover frequency for AN) = [mmoL of AN reacted selectively to DPU] × [mmoL of Pd(II) complex used]$^{-1}$ × h^{-1}. [d] 15 min. [e] PdNM = palladium-based nanostructural material was prepared in ethanol instead of water and raw solution of PdNM was introduced to the reactor (the volume of the solution was adjusted to obtain 0.056 mmoL of Pd). [f] 66 µL of 4MePy added.

3. Materials and Methods

3.1. Materials

Palladium chloride, sodium chloride, and sodium borohydride were used as received. Pyridine (Py), 2-methylpyridine (2-MePy), 3-methylpyridine (3-MePy), 4-methylpyridine (4-MePy), 2,6-dimethylpyridine (2,6-Me$_2$Py), 2,4-dimethylpyridine (2,4-Me$_2$Py), 3,5-dimethylpyridine (3,5-Me$_2$Py), 2-chloropyridine (2-ClPy), 3-chloropyridine (3-ClPy), 2,4-dichloropyridine (2,4-Cl$_2$Py), aniline, and ethanol were distilled (or fractionally distilled) over drying agent and stored under argon. 2,6-dichloropyridine (2,6-Cl$_2$Py), 3,5-dichloropyridine (3,5-Cl$_2$Py), iron powder, and iodine were used as received. Ultrapure (Milli-Q, 18.2 MΩ·cm resistivity at 25 °C) water was used in all experiments.

3.2. Synthesis of Palladium Nanoparticles

PdNPs were prepared according to the method described elsewhere [48,55]. NaCl (0.188 mmoL; 0.11 g) and PdCl$_2$ (0.084 mmoL; 0.015 g) were dissolved in 6 mL of ultrapure water and stirred at room temperature to form water soluble PdCl$_4^{2-}$ species. Then, a freshly prepared solution of ligand (derivative of pyridine; 0.628 mmoL in 9 mL of water) was added, stirred for 20 min, and reduced by NaBH$_4$ (1% w/v, 1.1 mL) added in 20 µL portions. The progress of reaction was observed as an immediate darkening of the mixture from light orange to dark brown (almost black). The resulting PdNPs were stirred for 30 min. In some experiments, for comparison purposes, nanoparticles were synthesized in ethanol instead of water. The size of centrifuged NPs as well as centrifuged and dried NPs was measured by transmission electron microscopy (TEM). The composition of Pd/ligand expressed as a percentage of the organic ligand and the metal in the centrifuged and dried palladium nanoparticles stabilized by 4-methylpyridine (PdNPs/4MePy) was determined by thermogravimetry (TG) under nitrogen atmosphere, with the heating rate = 10 K/min (see Figure S1 in Supplementary Information).

3.3. Techniques

Transmission electron microscopy (TEM) observations were carried out using JEM 1400 JEOL Co. microscope, at 120 kV acceleration voltage. The samples were obtained by casting aqueous (or ethanol) solution of palladium nanoparticles onto a carbon coated nickel microgrid (200 mesh) and air-dried overnight. The thermogravimetric (TG) measurements of PdNPs/4-MePy were performed with thermogravimeter Q50-1261 TA Instruments (USA) under nitrogen flow (6 dm^3/h), heating rate = 10 K/min. Weight loss during thermal decomposition of PdNPs/4-MePy was determined in the temperature range 40–600 °C. TG measurements were performed in platinum pan, and the weight of the sample was around 1–2 mg. Results presented in this paper are the arithmetic mean of three repetitions, and the difference of results in a series of determinations of the sample was up to 2%.

3.4. Carbonylation of Aniline by CO/CO$_2$/O$_2$ to N,N'-Diphenylurea

The procedure described elsewhere [15] was applied with some modifications. Briefly, the reaction was carried out in a 200 mL stainless-steel autoclave equipped with magnetic stirrer. Before the experiment, the autoclave was heated at 120 °C for 3 h (evaporation of water in order to avoid shifting the balance to the left, which may occur in the presence of an excess of water at high temperature) and cooled down to room temperature. Subsequently, one of the following catalysts: PdCl$_2$(X$_n$Py)$_2$, PdCl$_2$, Pd$_{black}$, or PdNPs (0.056 mmoL), and Fe powder (0–2.7 mmoL) were placed in the autoclave, the air was evacuated, and the system was filled with purified argon. Then, under a gentle stream of argon, other reagents and solvents were added: I$_2$ (0–0.39 mmoL), aniline (54 mmoL), ethanol (20 mL), and optionally Py or X$_n$Py (6.2 mmoL). After getting its cover closed, the autoclave was directly filled with molecular oxygen (0.6 MPa), a mixture of carbon dioxide and carbon monoxide (pressure of CO$_2$/CO = 3.4 MPa), then placed in a hot oil bath, and kept at 80–140 °C for 15 or 60 min, depending on the reaction. After 15 or 60 min (depending on the reaction), the autoclave was cooled in a water bath, and then vented. The solid phase obtained after centrifugation (15,000 rpm for 15 min) in the form of white needles (with traces of precipitated palladium black) was re-crystallized from methanol and analyzed by elemental analysis, IR and ^{1}H NMR. N,N-diphenylurea was identified on the basis of elemental analysis % (exp./calc.): C(73.60/73.57), H (5.75/5.70), N (13,22/13.20); FT-IR (KBr): 3326, 3283(s) ν_{NH}; 3000–3100 (s) $\nu_{C-H\,aromat}$; 1648 (s) $\nu_{C=O}$; 1595, 1555 (m) $\nu_{C=Caromat}$; 1496, 1444 (m) ν_{N-H}; 1313, 1233 (m) ν_{C-N}; 755, 697 (s) $\nu_{C-Haromat}$ cm^{-1}. M.P. = 235–237 °C; ^{1}H NMR (300 MHz, DMSO): δ (ppm): 8.65 (s, 2 H), 7.46 (d, 4 H), 7.28 (t, 4 H), 6.97 (t, 2 H), and obtained values are in agreement with literature data [67]. Analysis of the liquid phase was performed using gas chromatography (GC-FID, GC-MS). Calculation of conversion of aniline was based on GC-FID analysis with n-decyl alcohol as standard.

4. Conclusions

In this work, the effect of CO$_2$ on the rate of carbonylation of aniline to N,N'-diphenylurea was investigated. Taking into account results obtained for various types of Pd-based pre-catalysts—namely, PdNPs, PdNM, Pd(0) and Pd(II) complexes—the highest TOF is observed for PdNPs. Thus, we suggest that PdNPs act as active species or as the most efficient source of other catalytically active Pd(0) species. High yields observed for Pd(II) complexes support proposed mechanism where Pd(II) is easily reduced to catalytically active Pd(0) during catalytic cycle. We conclude that introduction of PdNPs to the reaction mixture instead of Pd$_{black}$ or Pd(II) results in higher conversion and turnover frequency. We believe that obtained results may be helpful in elucidating the role of metallic nanoparticles in other organic processes.

Supplementary Materials: The following are available online at http://www.mdpi.com/2073-4344/10/8/877/s1, Figure S1: TG curve for centrifuged and dried PdNPs/4MePy (obtained in water), Figure S2: Effect of derivatives of pyridine on the conversion (Conv.), yield and selectivity (Select.) of the catalyst. Reaction conditions: 54 mmol AN, PdCl$_2$(X$_n$Py)$_2$/Fe/I$_2$ = 0.056/2.68/0.118 mmol, 15 atm CO, 19 atm CO$_2$, 6 atm O$_2$, 20 ml EtOH, 100°C, 60 min.

Author Contributions: D.M.: Investigation, Visualization, Writing—original draft, Writing—review & editing. A.K.: Visualization, Formal analysis, Writing—review & editing. P.P.: Methodology, Data curation, Writing—original draft, Writing—review & editing. A.K.-S.: Conceptualization, Investigation, Data curation, Formal analysis, Funding acquisition, Methodology, Project administration, Resources, Supervision, Visualization, Writing—original draft, Writing—review & editing. All authors have read and agreed to the published version of the manuscript.

Funding: This work was supported by the National Centre for Research and Development-Poland, grant LIDER no. LIDER/34/0102/L-7/15/NCBR/2016.

Conflicts of Interest: This manuscript has not been submitted elsewhere for consideration, and the authors declare no competing financial interests.

References

1. Ferretti, F.; Barraco, E.; Gatti, C.; Ramadan, D.R.; Ragaini, F. Palladium/iodide catalyzed oxidative carbonylation of aniline to diphenylurea: Effect of ppm amounts of iron salts. *J. Catal.* **2019**, *369*, 257–266. [CrossRef]
2. Wu, X.-F.; Neumann, H.; Beller, M. Palladium-Catalyzed Oxidative Carbonylation Reactions. *ChemSusChem* **2013**, *6*, 229–241. [CrossRef] [PubMed]
3. Didgikar, M.R.; Roy, D.; Gupte, S.P.; Joshi, S.S.; Chaudhari, R.V. Immobilized Palladium Nanoparticles Catalyzed Oxidative Carbonylation of Amines. *Ind. Eng. Chem. Res.* **2010**, *49*, 1027–1032. [CrossRef]
4. Guan, Z.H.; Lei, H.; Chen, M.; Ren, Z.H.; Bai, Y.; Wang, Y.Y. Palladium-Catalyzed Carbonylation of Amines: Switchable Approaches to Carbamates and N,N'-Disubstituted Ureas. *Adv. Synth. Catal.* **2012**, *354*, 489–496. [CrossRef]
5. Gabriele, B.; Salerno, G.; Mancuso, R.; Costa, M. Efficient Synthesis of Ureas by Direct Palladium-Catalyzed Oxidative Carbonylation of Amines. *J. Org. Chem.* **2004**, *69*, 4741–4750. [CrossRef]
6. Katritzky, A.R.; Oliferenko, A.; Lomaka, A.; Karelson, M. Six-Membered cyclic ureas as HIV-1 protease inhibitors: A QSAR study based on CODESSA PRO approach. *Bioorg. Med. Chem. Lett.* **2002**, *12*, 3453–3457. [CrossRef]
7. Patel, M.; Rodgers, J.D.; McHugh, R.J., Jr.; Johnson, B.L.; Cordova, B.C.; Klabe, R.M.; Bacheler, L.T.; Erickson-Viitanen, S. Ko SS Unsymmetrical cyclic ureas as HIV-1 protease inhibitors: Novel biaryl indazoles as P2/P2' substituents. *Bioorg. Med. Chem. Lett.* **1999**, *9*, 3217–3220. [CrossRef]
8. Inaloo, I.D.; Majnooni, S.A. Fe_3O_4@SiO_2/Schiff Base/Pd Complex as an Efficient Heterogeneous and Recyclable Nanocatalyst for One-Pot Domino Synthesis of Carbamates and Unsymmetrical Ureas. *Eur. J. Org. Chem.* **2019**, *37*, 6359–6368. [CrossRef]
9. Klaus, S.; Lehenmeier, M.W.; Anderson, C.E.; Rieger, B. Recent advances in CO_2/epoxide copolymerization—New strategies and cooperative mechanisms. *Coord. Chem. Rev.* **2011**, *255*, 1460–1479. [CrossRef]
10. Mulla, S.A.R.; Rode, C.V.; Kelkar, A.A.; Gupte, S.P. Activity of homogeneous transition metal catalysts for oxidative carbonylation of aniline to N,N'diphenyl urea. *J. Mol. Catal. A Chem.* **1997**, *122*, 103–109. [CrossRef]
11. Bigi, F.; Maggi, R.; Sartori, G. Selected syntheses of ureas throuth phosgene substitutes. *Green Chem.* **2000**, *2*, 140–148. [CrossRef]
12. Zahrtmann, N.; Claver, C.; Godard, C.; Riisager, A.; Garcia-Suarez, E. Selective Oxidative Carbonylation of Aniline to Diphenylurea with Ionic Liquids. *J. ChemCatChem* **2018**, *10*, 2450–2457. [CrossRef]
13. Mancuso, R.; Raut, D.S.; Della Ca', N.; Fini, F.; Carfagna, C.; Gabriele, B. Catalytic Oxidative Carbonylation of Amino Moieties to Ureas, Oxamides, 2-Oxazolidinones, and Benzoxazolones. *ChemSusChem* **2015**, *13*, 2204–2211. [CrossRef] [PubMed]
14. Krogul, A.; Litwinienko, G. Application of Pd(II) Complexes with Pyridines as Catalysts for the Reduction of Aromatic Nitro Compounds by CO/H_2O. *Org. Process. Res. Dev.* **2015**, *19*, 2017–2021. [CrossRef]
15. Krogul, A.; Litwinienko, G. One pot synthesis of ureas and carbamates via oxidative carbonylation of aniline-type substrates by CO/O_2 mixture catalyzed by Pd-complexes. *J. Mol. Catal. A Chem.* **2015**, *407*, 204–211. [CrossRef]
16. Krogul, A.; Skupinska, J.; Litwinienko, G. Tuning of the catalytic properties of $PdCl_2(X_nPy)_2$ complexes by variation of the basicity of aromatic ligands. *J. Mol. Catal. A Chem.* **2014**, *385*, 141–148. [CrossRef]
17. Krogul, A.; Skupinska, J.; Litwinienko, G. Catalytic activity of $PdCl_2$ complexes with pyridines in nitrobenzene carbonylation. *J. Mol. Catal. A Chem.* **2011**, *337*, 9–16. [CrossRef]
18. Bartish, C.M.; Drissel, G.M. *Kirk-Othmer Encyclopedia of Chemical Technology*, 3rd ed.; Wiley-Interscience: New York, NY, USA, 1978; Volume 4, p. 774.
19. Cheng, J.; Zhou, F. Revised Explosibility Diagram to Judge Best Practice of Controlling an Explosive Gas-Mixture. *Fire Technol.* **2015**, *51*, 293–308. [CrossRef]

20. Chen, Y.; Hone, C.A.; Gutmann, B.; Kappe, C.O. Continuous Flow Synthesis of Carbonylated Heterocycles via Pd-Catalyzed Oxidative Carbonylation Using CO and O_2 at Elevated Temperatures and Pressures. *Org. Process Res. Dev.* **2017**, *21*, 1080–1087. [CrossRef]
21. Liu, A.H.; Li, Y.N.; He, L.N. Organic synthesis using carbon dioxide as phosgene-free carbonyl reagent. *Pure Appl. Chem.* **2012**, *84*, 581–602. [CrossRef]
22. Darensbourg, D.J. Making Plastics from Carbon Dioxide: Salen Metal Complexes as Catalysts for the Production of Polycarbonates from Epoxides and CO_2. *Chem. Rev.* **2007**, *107*, 2388–2410. [CrossRef] [PubMed]
23. Braunstein, P.; Matt, D.; Nobel, D. Reactions of Carbon Dioxide with Carbon-Carbon Bond Formation Catalyzed by Transition-Metal Complexes. *Chem. Rev.* **1988**, *88*, 747–764. [CrossRef]
24. Sakakura, T.; Choi, J.C.; Yasuda, H. Transformation of Carbon Dioxide. *Chem. Rev.* **2007**, *107*, 2365–2387. [CrossRef] [PubMed]
25. Kamphuis, A.J.; Picchioni, F.; Pescarmona, P.P. CO_2-fixation into cyclic and polymeric carbonates: Principles and applications. *Green Chem.* **2019**, *21*, 406–448. [CrossRef]
26. Shen, Y. Carbon dioxide bio-fixation and wastewater treatment *via* algae photochemical synthesis for biofuels production. *RSC Adv.* **2014**, *4*, 49672–49722. [CrossRef]
27. Leclaire, J.; Heldebrant, D.J. A call to (green) arms: A rallying cry for green chemistry and engineering for CO_2 capture, utilisation and storage. *Green Chem.* **2018**, *20*, 5058–5081. [CrossRef]
28. Liu, Q.; Wu, L.; Jackstell, R.; Beller, B. Using carbon dioxide as a building block in organic synthesis. *Nat. Commun.* **2015**, *6*, 5933. [CrossRef]
29. Wu, C.; Wang, J.; Chang, P.; Cheng, H.; Yu, Y.; Wu, Z.; Donga, D.; Zhao, F. Polyureas from diamines and carbon dioxide: Synthesis, structures and properties. *Phys. Chem. Chem. Phys.* **2012**, *14*, 464–468. [CrossRef]
30. Franz, M.; Stalling, T.; Steinert, H.; Martens, J. First catalyst-free CO_2 trapping of N-acyliminium ions under ambient conditions: Sustainable multicomponent synthesis of thia- and oxazolidinyl carbamates. *Org. Biomol. Chem.* **2018**, *16*, 8292–8304. [CrossRef]
31. Krawczyk, T.; Jasiak, K.; Kokolus, A.; Baj, S. Polymer- and Carbon Nanotube-Supported Heterogeneous Catalysts for the Synthesis of Carbamates from Halides, Amines, and CO_2. *Catal. Lett.* **2016**, *146*, 1163–1168. [CrossRef]
32. Nomura, R.; Hasegawa, Y.; Ishimoto, M.; Toyasaki, T.; Matauda, H. Carbonylation of Amines by Carbon Dioxide in the Presence of an Organoantimony Catalyst. *J. Org. Chem.* **1992**, *57*, 7339–7342. [CrossRef]
33. Dibenedetto, A.; Aresta, M.; Narracci, M. Carbon Dioxide Capture by Amines: Increasing the Efficiency by Amine Structure Modification. *Fuel Chem. Div. Prep.* **2002**, *47*, 53.
34. Abla, M.; Choi, J.C.; Sakakura, T. Halogen-free process for the conversion of carbon dioxide to urethanes by homogeneous catalysis. *Chem. Commun.* **2001**, 2238–2239. [CrossRef] [PubMed]
35. Heyn, R.H.; Jacobs, I.; Carr, R.H. Synthesis of Aromatic Carbamates from CO_2: Implications for the Polyurethane Industry. *Adv. Inorg. Chem.* **2014**, *66*, 83–115.
36. Choi, J.C.; Yuan, H.Y.; Fukaya, N.; Onozawa, S.; Zhang, Q.; Choi, S.J.; Yasuda, H. Halogen-Free Synthesis of Carbamates from CO_2 and Amines Using Titanium Alkoxides. *Chem. Asian J.* **2017**, *12*, 1297–1300. [CrossRef]
37. Zhang, Q.; Yuan, H.Y.; Fukaya, N.; Choi, J.C. Alkali Metal Salt as Catalyst for Direct Synthesis of Carbamate from Carbon Dioxide. *ACS Sustain. Chem. Eng.* **2018**, *6*, 6675–6681. [CrossRef]
38. Ren, Y.; Rousseaux, S.A.L. Metal-Free Synthesis of Unsymmetrical Ureas and Carbamates from CO_2 and Amines via Isocyanate Intermediates. *J. Org. Chem.* **2018**, *83*, 913–920. [CrossRef] [PubMed]
39. Inaloo, I.D.; Majnoonib, S. Carbon dioxide utilization in the efficient synthesis of carbamates by deep eutectic solvents (DES) as green and attractive solvent/catalyst Systems. *New J. Chem.* **2019**, *43*, 11275–11281. [CrossRef]
40. Lam, R.H.; McQueen, C.M.A.; Pernik, I.; McBurney, R.T.; Hill, A.F.; Messerle, B.A. Selective formylation or methylation of amines using carbon dioxide catalysed by a rhodium perimidine-based NHC complex. *Green Chem.* **2019**, *21*, 538–549. [CrossRef]
41. Riemer, D.; Hirapara, P.; Das, S. Chemoselective Synthesis of Carbamates Using CO_2 as Carbon Source. *ChemSusChem* **2016**, *9*, 1916–1920. [CrossRef]
42. Wang, P.; Fei, Y.; Deng, Y. Transformation of CO_2 into polyureas with 3-amino-1,2,4-triazole potassium as a solid base catalyst. *New J. Chem.* **2018**, *42*, 1202–1207. [CrossRef]

43. Honda, M.; Sonehara, S.; Yasuda, H.; Nakagawaa, Y.; Tomishige, K. Heterogeneous CeO_2 catalyst for the one-pot synthesis of organic carbamates from amines, CO_2 and alcohols. *Green Chem.* **2011**, *13*, 3406–3413. [CrossRef]
44. Ca', N.D.; Bottarelli, P.; Dibenedetto, A.; Aresta, M.; Gabriele, B.; Salerno, G.; Costa, M. Palladium-catalyzed synthesis of symmetrical urea derivatives by oxidative carbonylation of primary amines in carbon dioxide medium. *J. Catal.* **2011**, *282*, 120–127. [CrossRef]
45. Ma, L.; Xiao, Y.; Deng, J.; Wang, Q. Effect of CO_2 on explosion limits of flammable gases in goafs. *Min. Sci. Technol.* **2010**, *20*, 193–197. [CrossRef]
46. Leitner, W. Supercritical Carbon Dioxide as a Green Reaction Medium for Catalysis. *Acc. Chem. Res.* **2002**, *35*, 746–756. [CrossRef] [PubMed]
47. Subramaniam, B. Gas Expanded Liquids for Sustainable Catalysis. *Encycl. Sustain. Sci. Technol.* **2012**, 3933.
48. Flanagan, K.A.; Sullivan, J.A.; Mueller-Bunz, H. Preparation and Characterization of 4-Dimethylaminopyridine-Stabilized Palladium Nanoparticles. *Langmuir* **2007**, *23*, 12508–12520. [CrossRef]
49. Shaughnessy, K.H.; DeVasher, R.B. Palladium-Catalyzed Cross-Coupling in Aqueous Media: Recent Progress and Current Applications. *Curr. Org. Chem.* **2005**, *9*, 585–604. [CrossRef]
50. Vasylyev, M.V.; Maayan, G.; Hovav, Y.; Haimov, A.; Neumann, R. Palladium Nanoparticles Stabilized by Alkylated Polyetheneimine as Aqueous Biphasic Catalysts for the Chemoselective Stereocontrolled Hydrogenation of Alkenes. *Org. Lett.* **2006**, *8*, 5445–5448. [CrossRef]
51. Rucareanu, S.; Gandubert, V.J.; Lennox, R.B. 4-(N,N-Dimethylamino)pyridine-Protected Au Nanoparticles: Versatile Precursors for Water- and Organic-Soluble Gold Nanoparticles. *Chem. Mater.* **2006**, *18*, 4674–4680. [CrossRef]
52. Kaim, A.; Szydłowska, J.; Piotrowski, P.; Megiel, E. One-pot synthesis of gold nanoparticles densely coated with nitroxide spins. *Polyhedron* **2012**, *46*, 119–123. [CrossRef]
53. Oh, S.K.; Niu, Y.; Crooks, R.M. Size-Selective Catalytic Activity of Pd Nanoparticles Encapsulated within End-Group Functionalized Dendrimers. *Langmuir* **2005**, *21*, 10209–10213. [CrossRef] [PubMed]
54. Zhao, M.; Crooks, R.M. Homogeneous Hydrogenation Catalysis with Monodisperse, Dendrimer-Encapsulated Pd and Pt Nanoparticles. *Angew. Chem. Int. Ed.* **1999**, *38*, 364–366. [CrossRef]
55. Krogul-Sobczak, A.; Kasperska, P.; Litwinienko, G. N-heterocyclic monodentate ligands as stabilizing agents for catalytically active Pd-nanoparticles. *Catal. Commun.* **2018**, *104*, 86–90. [CrossRef]
56. Krogul-Sobczak, A.; Cedrowski, J.; Kasperska, P.; Litwinienko, G. Reduction of Nitrobenzene to Aniline by CO/H_2O in the Presence of Palladium Nanoparticles. *Catalysts* **2019**, *9*, 404. [CrossRef]
57. Ragaini, F.; Cenini, S. Mechanistic studies of palladium-catalysed carbonylation reactions of nitro compounds to isocyanates, carbamates and ureas. *J. Mol. Catal. A Chem.* **1996**, *109*, 1–25. [CrossRef]
58. Stahl, S.S. Palladium Oxidase Catalysis: Selective Oxidation of Organic Chemicals by Direct Dioxygen-Coupled Turnover. *Angew. Chem. Int. Ed.* **2004**, *43*, 3400–3420. [CrossRef]
59. Fukuoka, S.; Chono, M.; Kohno, M. Isocyanate without phosgene. *Chemtech* **1984**, *14*, 670–676.
60. Gupte, S.P.; Chaudhari, R.V. Oxidative carbonylation of aniline over PdC catalyst: Effect of promoters, solvents, and reaction conditions. *J. Catal.* **1988**, *114*, 246–258. [CrossRef]
61. Pri-Bar, I.; Schwartz, J. I_2-Promoted Palladium-Catalyzed Carbonylation of Amines. *J. Org. Chem.* **1995**, *60*, 8124–8125. [CrossRef]
62. Shi, F.; Deng, Y.; SiMa, T.; Yang, H. A novel ZrO_2-SO_4^{2-} supported palladium catalyst for syntheses of disubstituted ureas from amines by oxidative carbonylation. *Tetrahedron Lett.* **2001**, *42*, 2161–2163. [CrossRef]
63. McCusker, J.E.; Qian, F.; McElwee-White, L. Catalytic oxidative carbonylation of aliphatic secondary amines to tetrasubstituted ureas. *J. Mol. Catal. A Chem.* **2000**, *159*, 11–17. [CrossRef]
64. Maitlis, P.M.; Haynes, A.; James, B.R.; Catellani, M.; Chiusoli, G.P. Iodide effects in transition metal catalyzed reactions. *Dalton Trans.* **2004**, 3409–3419. [CrossRef] [PubMed]
65. Gabriele, B.; Mancuso, R.; Veltri, L.; Della Ca', N. Polemic against conclusions drawn in "Palladium/iodide catalyzed oxidative carbonylation of aniline to diphenylurea: Effect of ppm amounts of iron salts". *J. Catal.* **2019**, *380*, 387–390. [CrossRef]

66. Benesovsky, F. *Gmelin Handbuch der Anorganischen Chemie*; Springer: New York, NY, USA, 1979; Volume 59, p. 337.
67. Kwiatkowski, A.; Jędrzejewska, B.; Józefowicz, M.; Grela, I.; Ośmiałowski, B. The trans/cis photoisomerization in hydrogen bonded complexes with stability controlled by substituent effects: 3-(6-aminopyridin-3-yl) acrylate case study. *RSC Adv.* **2018**, *8*, 23698–23710. [CrossRef]

© 2020 by the authors. Licensee MDPI, Basel, Switzerland. This article is an open access article distributed under the terms and conditions of the Creative Commons Attribution (CC BY) license (http://creativecommons.org/licenses/by/4.0/).

Article

Nickel-Modified TS-1 Catalyzed the Ammoximation of Methyl Ethyl Ketone

Dandan Yang, Haiyan Wang, Wenhua Wang, Sihua Peng, Xiuzhen Yang, Xingliang Xu * and Shouhua Jia *

College of Chemistry and Material Science, Shandong Agricultural University, Tai'an 271018, China; 2017110585@sdau.edu.cn (D.Y.); why15053865003@163.com (H.W.); w17863807408@163.com (W.W.); psh18865383970@163.com (S.P.); x976360872@163.com (X.Y.)
* Correspondence: xxlsdau@163.com (X.X.); shjia@sdau.edu.cn (S.J.); Tel.: +86-0538-824-1570 (S.J.)

Received: 5 November 2019; Accepted: 29 November 2019; Published: 4 December 2019

Abstract: In this paper, five kinds of transition metal-modified titanium silicalite-1 (M-TS-1) were prepared by an ultrasonic impregnation method. We studied their catalytic performances in the ammoximation of methyl ethyl ketone (MEK). The various M-TS-1 catalysts revealed distinct differences in their MEK ammoximation activity. The nickel-modified TS-1 (Ni-TS-1), especially 3 wt % Ni-TS-1, exhibited a satisfactory conversion of MEK (99%) associated with a high selectivity of methyl ethyl ketoxime (MEKO) (99.3%), which was 6% higher than that of TS-1 under the same conditions. Moreover, the catalyst showed excellent recyclability and the reactivity could be completely recovered after regeneration. The catalysts were characterized by Powder X-ray Diffraction (XRD), Fourier Transformed Infrared Spectra (FT-IR), X-ray photoelectron spectroscopy (XPS), and so on. It was demonstrated that the skeleton structure of TS-1 was basically maintained and the electron environment of the Ti active site was changed after the nickel modification, which can optimize the adsorption capacity and the activation for H_2O_2. Meanwhile, the surface nickel species reduced the surface acidity of the catalyst, which provided an appropriate pH and inhibited the deep oxidation of oxime.

Keywords: TS-1; nickel-modified; methyl ethyl ketone; ammoximation; methyl ethyl ketoxime

1. Introduction

Oxime is an important chemical raw material and intermediate, which is widely used in the synthesis of various high value-added chemicals. For example, cyclohexanone oxime is a key intermediate in the production of caprolactam as nylon-6 monomer [1], and methyl ethyl ketoxime (MEKO) can also be used as an important raw material to synthesize silicone crosslinkers, silicon curing agents, and the blocking agents of isocyanate, etc. [2–4]. The traditional synthesis method of oxime is the hydroxylamine method, which is a convenient and valuable method and a non-catalytic oximation of ketone with hydroxylamine derivative like $(NH_2OH)_2 \cdot H_2SO_4$. Unfortunately, the hydroxylamine method has many drawbacks, such as using toxic and highly acidic reagents, like hydroxylamine and sulfuric acid, while producing a large number of low-value by-products such as ammonium sulfate [5]. Compared with the traditional hydroxylamine method, the ammoximation with titanium silicate molecular sieve as catalyst and H_2O_2 as oxidant is a new method for the preparation of ketoxime which has been developed in recent years. Based on the concept of "green chemistry", the method has a series of advantages such as high efficiency, mild reaction conditions, atom economy, and only water as by-product (Scheme 1) [6,7].

Since Taramasso first synthesized titanium silicalite-1 (TS-1) [8], the development of titanium silicalite molecular sieve as a catalyst has been widely concerned. The active center of TS-1 with MFI topology is Ti^{4+} in the framework of molecular sieve. Because titanium oxygen tetrahedron is unstable, it is difficult to exist in the perfect form of tetra-coordination. It has the tendency to form six-coordination [9]. This means that the tetra-coordinated Ti^{4+} has electronic defects and the potential of accepting electron pairs. Therefore, it has unique adsorption and activation properties for H_2O_2 and catalyzes the selective oxidation of a variety of organic compounds [10]. At present, the ammoximation of cyclohexanone catalyzed by TS-1 has been industrialized. The conversion of cyclohexanone reached 99.9%, and the selectivity of cyclohexanone oxime was higher than 98.2% [11]. However, when the TS-1/H_2O_2 system was applied to the ammoximation of small ketones like methyl ethyl ketone (MEK), the selectivity of MEKO was not satisfactory. Under the same reaction conditions with cyclohexanone, the selectivity of MEKO is only 80%. The selectivity of MEKO can only reach about 95% by optimizing the conditions. The reason may be that the linear small ketoxime is easy to enter into the pore of the catalyst and is oxidized deeply [4].

$$R_1=O + NH_3 + H_2O_2 \xrightarrow{Catalyst} R_1=NOH + 2H_2O$$

Scheme 1. Preparation of oxime by the ammoximation of ketone.

Regulating electrophilicity of TS-1 to optimize adsorption ability and the Lewis acidic strength is a very effective strategy for preparing high-efficiency TS-1-based catalysts and improving the applications of TS-1 in oxidation reaction [12–14]. According to the catalytic properties of TS-1 and the mechanism of ammoximation [15], the modification of TS-1 to regulate its electrophilicity may be important to improve the selectivity of ketoxime in the ammoximation of ketone.

In recent years, transition metals (such as Ni, Fe, Co, Cu, etc.)-based catalysts have been extensively studied and applied in catalytic oxidation [16–19]. Some transition metal-modified TS-1 were used in the epoxidation of olefin. Wu et al. investigated the effect of transition metal (such as V, Cr, Mn, Fe, Co, Ni, Cu, Zn, Cd, La, 1% metals loading)-modified TS-1 on the epoxidation of butadiene. They found that Fe, Co, Ni can promote H_2O_2 conversion effectively and increase the electrophilicity of TS-1. These catalysts exhibit significant enhancement in butadiene selective epoxidation. However, Cu can inhibit H_2O_2 conversion in this reaction, and the interaction between Cu and Ti is relatively weak [20,21]. Capel-Sanchez et al. carried out the epoxidation of 1-octene with the TS-1 modified by several metal ions (Li^+, Ca^{2+}, La^{3+}, and Ce^{4+}). They found that the addition of metal cations could significantly improve the selectivity of epoxides, which is attributed to the addition of metal oxides neutralizing the surface acidity of TS-1 zeolite, and thus inhibiting the solvolysis reaction of epoxides at acidic sites and improving the selectivity of epoxides [22]. Nevertheless, there are few reports about transition metal-modified TS-1 used in the ammoximation of ketone.

Herein, we prepared a series of transition metal-modified TS-1 catalysts by an ultrasonic impregnation method. In comparison with the existing catalysts, the nickel-modified TS-1 catalysts, especially 3 wt % Ni-TS-1, showed great improvements in the ammoximation of MEK to synthesize MEKO with H_2O_2 as the oxidant. A detailed characterization and the effects of various experimental parameters were systematically conducted and investigated, where the nature of the promoted catalytic performances in the ammoximation of MEK was revealed. Moreover, the catalyst has successfully recovered without considerable loss of MEK conversion and MEKO selectivity.

2. Results and Discussion

2.1. Catalytic Activity on the Ammoximation of MEK

2.1.1. The Catalytic Activity of M-TS-1

Five kinds of transition metal (Fe, Co, Ni, Cu, Ce)-modified TS-1 catalysts were prepared. Moreover, the catalytic performances of M-TS-1 on the ammoximation of MEK were studied. The results are shown in Table 1. It is apparent that the various M-TS-1 catalysts revealed distinct different catalytic activity. The catalytic effect of Cu-TS-1 was the worst. The conversion of MEK was reduced from 98.9%

of TS-1 to 74.1%. The selectivity of MEKO was 0, that is, MEK was all converted into by-products. Although the Cu can interact with Ti active sites [20], it has little catalytic effect in the ammoximation of MEK. This may be explained that copper may catalyze H_2O_2 to produce highly active free radicals [23], effectively oxidizing and decomposing MEK and MEKO. In the ammoximation of MEK catalyzed by Fe-TS-1 and Co-TS-1, the conversion of MEK was still lower than that of TS-1, which may be attributed to the oxides of iron or cobalt that can decompose H_2O_2, which is not conducive to the ammoximation of MEK. For the introduction of Fe, the selectivity of MEKO slightly improved from 92.8% of TS-1 to 95.6%. It confirmed that the electronic effect of Fe on Ti centers would activate TS-1 activity [21]. When the rare earth element cerium was introduced, the conversion of MEK was 97.3%, which was similar to TS-1, probably because the cerium's atomic radius was too large to recombine with TS-1. It was worth noting that Ni-TS-1 displayed the best catalytic activity in the conversion of MEK to MEKO. Relative to TS-1, the selectivity of MEKO (99.3%) was significantly improved while maintaining high conversion of MEK (99%). Thus, Ni-TS-1 was intensively studied in the following research.

Table 1. Comparisons of the catalytic performance for the ammoximation of methyl ethyl ketone (MEK) over various catalysts.

Catalyst	Conversion (%)	Selectivity (%)
TS-1 [a]	98.9	92.8
Fe-TS-1 [a]	96.5	95.6
Co-TS-1 [b]	90.7	91.3
Ni-TS-1 [a]	99	99.3
Cu-TS-1 [b]	74.1	0
Ce-TS-1 [a]	97.3	90.6

[a] Reaction conditions: MEK, 0.1 mol; t-butanol, 25 mL; catalyst, 1.00 g; temperature, 343 K; total reaction time, 2 h. The H_2O_2 and $NH_3 \cdot H_2O$ were added at a constant rate for 1.5 h. [b] Total reaction time, 3.5 h. The H_2O_2 and $NH_3 \cdot H_2O$ were added at a constant rate for 1.5 h. The other conditions were the same with [a].

2.1.2. The Catalytic Activity of XNi-TS-1

In order to optimize the catalytic activity of Ni-TS-1, the effects of Ni dosage in Ni-TS-1 on the ammoximation of MEK was investigated. The results are shown in Table 2. The selectivity of MEKO went up and then down gradually with the increase of Ni dosage. When the dosage was 3 wt %, the catalytic activity was the best. This apparent change may be associated with the structure of the catalyst. When the Ni dosage was small, nickel species did not affect the Ti active sites. When the Ni dosage was too high, there were many Ni species to block the channels and cover Ti active sites, thus it was adverse to catalytic reactions.

Table 2. The effects of Ni dosage on the ammoximation of MEK over Ni-titanium silicalite-1 (TS-1).

Catalyst	Conversion (%)	Selectivity (%)
1 wt % Ni-TS-1	99.0	92.4
2 wt % Ni-TS-1	98.8	93.9
3 wt % Ni-TS-1	99.0	99.3
4 wt % Ni-TS-1	99.1	96.7
5 wt % Ni-TS-1	99.1	91.4

Reaction conditions: MEK, 0.1 mol; t-butanol, 25 mL; catalyst, 1.00 g; reaction temperature, 343 K; total reaction time, 2 h. The H_2O_2 and $NH_3 \cdot H_2O$ were added at a constant rate for 1.5 h.

2.2. Reusability Tests of 3% Ni-TS-1

The main advantages of catalysts in heterogeneous catalytic systems are its good stability and recyclability. The 3 wt % Ni-TS-1 catalyst was reused and its service life was tested. As can be seen from Figure 1, the conversion of MEK and the selectivity of MEKO were not affected when used twice. When the catalyst was reused five times, the conversion of MEK remained above 90%, and the selectivity of MEKO was over 74.1%. The decrease of catalytic activity may be attributed to the loss of a small number of catalysts, increase of surface acidity, and the obstruction of active sites by organic species. The catalyst after each reuse was dried. Then thermogravimetric analysis was carried out, and the weight loss rate was calculated (Table 3). It was found that the weight loss rate increased with the number of uses, and the blockage of pore channels by organic substances became more and more serious. This may be the main factor affecting the catalytic activity. After using five times, 3 wt % Ni-TS-1 was regenerated by calcination and applied to MEK ammoximation reaction (Figure 1). The catalytic activity of 3 wt % Ni-TS-1 was completely restored.

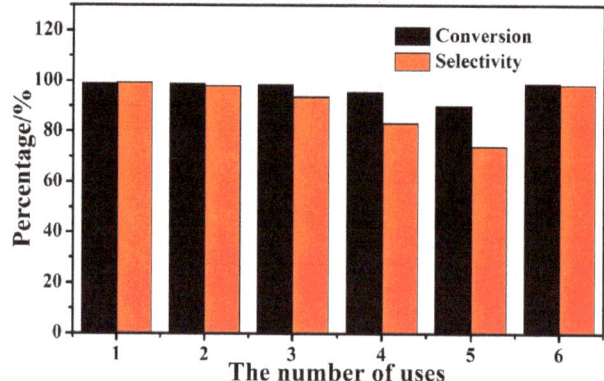

Figure 1. Repetition and regeneration of 3 wt % Ni-TS-1 on the ammoximation of MEK (Reaction conditions: MEK, 0.1 mol; t-butanol, 25 mL; catalyst, 1.00 g; reaction temperature, 343 K; total time, 2 h. The total time of other reactions were 3.5 h. The H_2O_2 and $NH_3 \cdot H_2O$ were added at a constant rate for 1.5 h).

Table 3. Thermogravimetric analysis after repeated use of 3 wt % Ni-TS-1.

The Number of Uses	Weight Loss Rate (%)
1	5.10661
2	8.21042
3	8.90195
4	9.68479
5	10.20894

2.3. Characterization of Catalysts

2.3.1. XRD and FT-IR Characterization

The crystal structure information of commercial TS-1, M-TS-1, and XNi-TS-1 were analyzed by the XRD (Figure 2A,B). All samples exhibited same diffraction peaks at 2θ = 7.8°, 8.8°, 23.2°, 23.8°, and 24.3°, which represented that the MFI skeleton structure of TS-1 was not changed or destroyed after metal modification. There was no characteristic diffraction peak of metal oxides in the XRD patterns. It may be that parts of metal entered the framework of TS-1 and a small amount of metal oxides were well dispersed on the TS-1 surface by the ultrasonic impregnation synthesis method [24].

The stretching vibration of chemical bonds or functional groups in the TS-1 and M-TS-1 samples were detected by FT-IR spectroscopy (Figure 2C). The vibration peaks at 550 cm^{-1} and 1225 cm^{-1} were characteristic peaks of the molecular sieve with MFI topology; 450 cm^{-1} was the bending vibration peak of Si–O bond; 1110 cm^{-1} and 800 cm^{-1} absorption peaks corresponded to antisymmetric and symmetrical stretching vibration of internal silicon-oxygen tetrahedral units [25]; the characteristic peak of 960 cm^{-1} indicated the existence of the tetra-coordinated framework titanium. After the metals modified TS-1, the characteristic peak of 960 cm^{-1} still existed obviously. However, the modification of metals made the infrared absorption band shift at around 966 cm^{-1}. In the different metals used to modify the TS-1 catalyst, the metal with the greatest wave number migration is Ni, followed by Fe, Co, and Ce. With the addition of Cu, wave number migration is almost not realized. The local FT-IR spectrum of XNi-TS-1 is shown in Figure 2D. It can be seen that the introduction of Ni caused a deviation in the IR absorption bands around 965 cm^{-1} to lower wave numbers. The wave number of [Ti–O] bond vibration decreased from 965.37 cm^{-1} to 962.11 cm^{-1} as the dosage of Ni increased from 1 wt % to 5 wt %, which indicated that the introduction of Ni weakened the [Ti–O] bond. It may be related to the charge transfer of [Ti–O] in the tetrahedral [26]. In other words, the interactions between Ni and Ti decreased the strength of the titanium oxygen bond and increased the electrophilicity of the Ti center as the dosage of Ni in the TS-1 increased, which made a substantial contribution to the satisfactory conversion of MEK to MEKO.

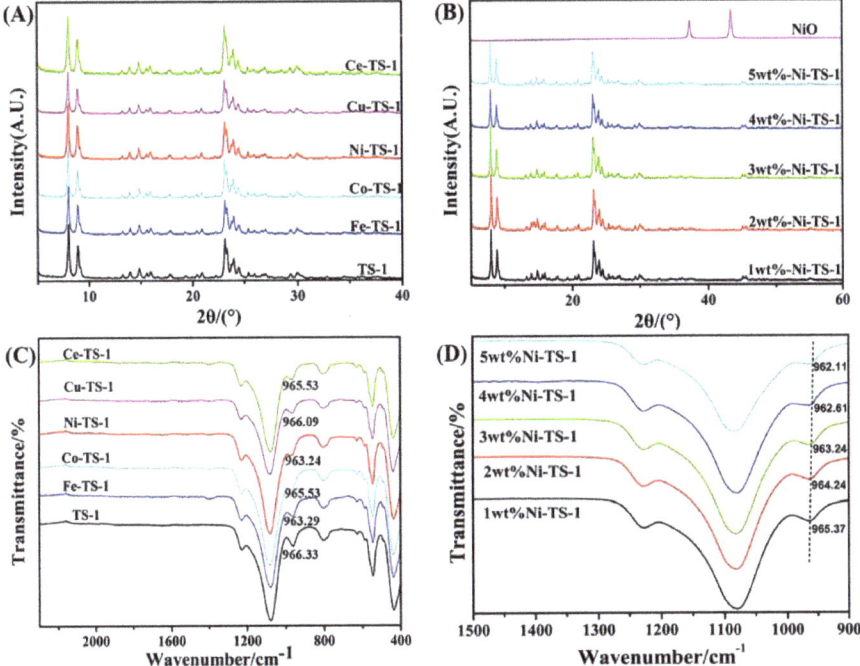

Figure 2. (**A**) XRD patterns of M-TS-1, (**B**) XRD patterns of XNi-TS-1, (**C**) FI-IR spectra of M-TS-1, and (**D**) FI-IR spectra of XNi-TS-1.

2.3.2. SEM Characterization

Microstructure of samples was further monitored by SEM. As can be seen from Figure 3, both TS-1 and Ni-TS-1 samples showed uniform spherical particles with the size of about 0.1–0.2 µm. Compared with TS-1, the size and shape of Ni-TS-1 did not have an obvious change. These results showed that the metal oxide did not destroy the crystal structure of TS-1, nor did it change the morphology of TS-1 particles.

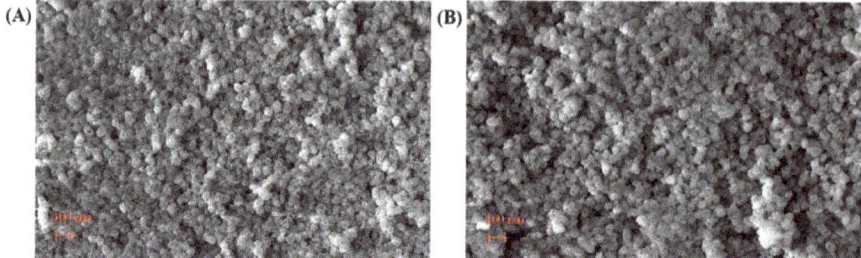

Figure 3. SEM images of TS-1 (**A**) and 3 wt % Ni-TS-1 (**B**).

2.3.3. Induced Coupled Plasma-Atomic Emission Spectroscopy (ICP-AES) and N_2 Adsorption-Desorption Characterization

The results of ICP-AES for the XNi-TS-1 catalyst are shown in Table 4. The loading efficiency (e) of Ni on TS-1 was about 70%.

Table 4. Ni content, specific surface area, and micropore volume of TS-1 and XNi-TS-1.

Catalyst	w (Ni)(%) Actual	e (Ni)(%)	BET Surface area/$m^2 \cdot g^{-1}$	Micropore Volume/$cm^3 \cdot g^{-1}$
TS-1	-	-	394.7	0.0722
1 wt % Ni-TS-1	0.73	73.0	388.0	0.0719
2 wt % Ni-TS-1	1.39	69.5	387.9	0.0714
3 wt % Ni-TS-1	2.07	69.0	384.9	0.0709
4 wt % Ni-TS-1	2.75	68.8	377.9	0.0708
5 wt % Ni-TS-1	3.43	68.6	363.3	0.0699

N_2 adsorption–desorption isotherm of TS-1 and XNi-TS-1 are shown in Figure 4. According to the BDDT classification, all of the samples showed type IV isotherms with type H3 hysteresis loop, indicating the presence of mesopores. In addition, the introduction of Ni did not affect the presence of mesopores in TS-1. As shown in Table 4, the decrease in surface area and pore volume can be observed with the increase of Ni dosage. The N_2 adsorption capacity of the Ni-TS-1 decreased compared to TS-1. This suggested that the metal oxides partially occupied the microporous channels.

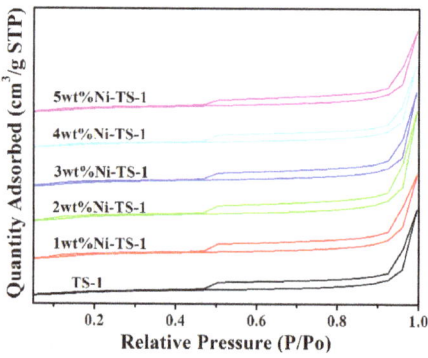

Figure 4. N_2 adsorption–desorption isotherm of TS-1 and XNi-TS-1.

2.3.4. DR UV-Vis Characterization

DR UV-Vis spectroscopy is an effective way to understand the nature of active Ti species on TS-1. The DR UV–Vis spectroscopy results of TS-1 and XNi-TS-1 are presented in Figure 5. In the DR UV-Vis spectra, the absorption peak near 210 nm indicated the existence of tetra-coordinated skeleton titanium, which was originated from the electronic transfer of the pπ–pπ transitions between titanium and oxygen in the framework titanium species. The absorption peaks around 330 nm was anatase TiO_2 outside of the framework [27–29]. It can be seen from Figure 5 that the absorption peak of TS-1 near 330 nm was large, and the peak of Ni-TS-1 still existed, but there was a tendency to become smaller from 1 wt % Ni-TS-1 to 5 wt % Ni-TS-1. This may be due to the addition of Ni, which caused some non-skeletal titanium to be lost. In addition, no characteristic band at around 362 nm is observed in XNi-TS-1 samples, which is assigned to NiO. This conclusion agreed with the XRD results.

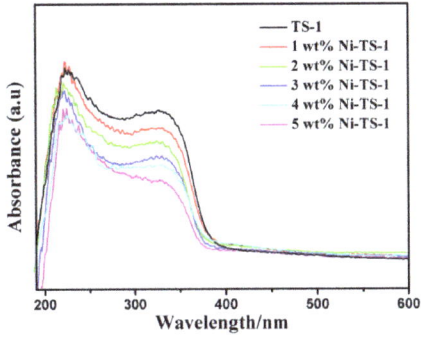

Figure 5. DR UV-Vis spectra of TS-1 and XNi-TS-1.

2.3.5. XPS Characterization

The surface chemical status and the interaction between the nickel component and TS-1 were investigated by XPS. Figure 6A shows the XPS full spectra of 3 wt % Ni-TS-1. Obviously, 3 wt % Ni-TS-1 had peaks of titanium, nickel, and silicon. The spectra of Ni 2p in 3 wt % Ni-TS-1 were fitted into four peaks in Figure 6B. The symbolic peaks at 856.5 eV and 874.2 eV are attributed to the binding energies of Ni $2p_{3/2}$ and Ni $2p_{1/2}$, which were assigned to Ni^{2+}. The Ni $2p_{3/2}$ XPS spectrum of free NiO usually shows peaks at 855.6 eV [30], while the Ni $2p_{3/2}$ XPS spectrum of 3 wt % Ni-TS-1 showed peaks at 856.5 eV. The increase of binding energy values indicated that Ni species in modified samples was afforded electrons by TS-1. The two satellite peaks, at a binding energy of 862.3 eV and 880.1 eV, for Ni^{2+} were also observed [31]. Figure 6C shows the Ti 2p spectra of TS-1 and 3 wt % TS-1. The two characteristic peaks located at 460.2 eV and 465.2 eV were attributed to Ti $2p_{3/2}$ and Ti $2p_{1/2}$, respectively [20]. The peak of Ti $2p_{3/2}$ could be decomposed into two components. The peaks at 458.7 eV and 460.2 eV were assigned to anatase TiO_2 and framework Ti species, respectively [32]. For nickel-modified TS-1 catalyst, in addition to the characteristic peak at 460.5 eV, another characteristic peak appeared at 459.7 eV. The decrease of binding energy may be due to the migration of Ti 2p orbital electron cloud in the skeleton, which reduced the density of the electron cloud around the Ti center. Combined with Ni 2p XPS results, the addition of Ni reduced the density of electron cloud around the Ti center in TS-1, and enhanced the electrophilicity of the Ti center, which can improve the catalytic activity of TS-1 on the selective catalytic oxidation reaction.

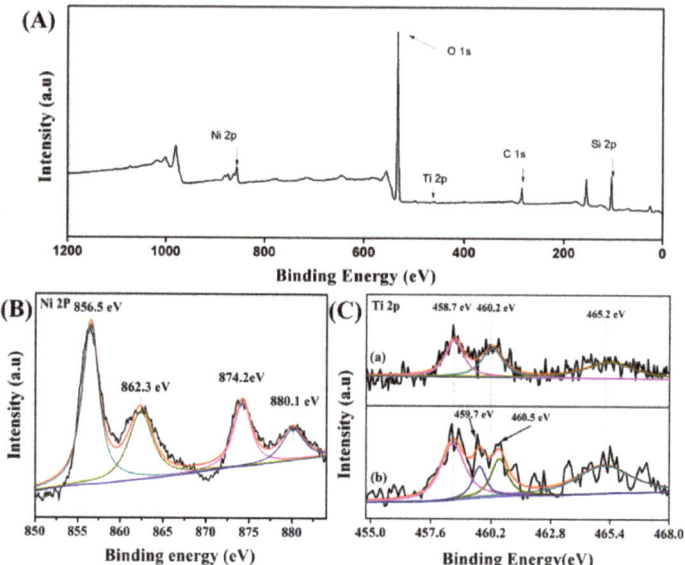

Figure 6. (A) XPS full spectra of 3 wt % Ni-TS-1, (B) Ni 2p, and (C) Ti 2p XPS of (a) TS-1 and (b) 3 wt % Ni-TS-1.

2.3.6. The Analysis of Point of Zero Charge

Point of zero charge (PZC) is the pH when the net charge on the solid surface is zero in the aqueous solution. It is an important parameter for calibrating acidity and basicity of a solid surface [33]. The PZC of TS-1 and 3 wt % Ni-TS-1 is shown in Figure 7. The PZC of TS-1 and 3 wt % Ni-TS-1 was 3.00 and 6.25, respectively. When TS-1 was modified by nickel, the PZC of the catalyst was raised. This striking observation could be closely related to the structure and characteristic of the sample. Firstly, Ni replaced Si into the framework of TS-1, because the PZC of NiO and SiO_2 is 8.33 and 3.00, respectively [34,35]. In addition, the acidic sites on the surface were covered by nickel species. The rise of PZC was beneficial to maintaining higher pH of ammoximation system of MEK and inhibiting the further oxidation of MEKO.

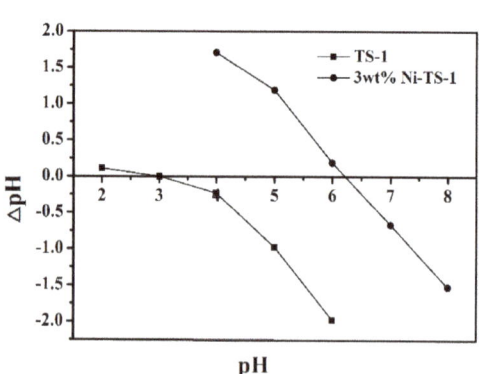

Figure 7. The point of zero charge (PZC) of TS-1 and 3 wt % Ni-TS-1.

2.4. Mechanism Analysis

The results of characterization illustrated parts of modified Ni were incorporated into the TS-1 skeleton as replacements of some Si sites and interacted with Ti active site and improved the electrophilicity of the Ti center [26], which is beneficial to adsorb H_2O_2 in the reaction system. the mechanism of liquid-phase ketoammoximation is hydroxylamine mechanism [36]. According to literature reports and our experimental results, we proposed the mechanism of Ni-TS-1 catalyzed the ammoximation of MEK (Scheme 2). Firstly, the four-coordinated titanium species combines with $NH_3·H_2O$ to become the six-coordinated state, further forming titanium peroxide under the action of H_2O_2. Then titanium peroxide reacts with $NH_3·H_2O$ to form ammonium peroxide, which would easily release NH_2OH. Eventually, NH_2OH would react with MEK to produce MEKO by non-catalytic oxidation [37]. The addition of Ni reduced the electron cloud density of the titanium active center, improved its electrophilicity, enhanced the adsorption capacity for H_2O_2, and thus increased the reaction rate of the control step of ketoammoximation, i.e., the formation rate of hydroxylamine. In addition, the increase of basicity of the Ni-modified catalyst was beneficial to controlling the alkaline environment of the reaction system and effectively inhibiting the occurrence of side reactions. Therefore, Ni-TS-1 exhibited a good catalytic effect in the ammoximation of MEK.

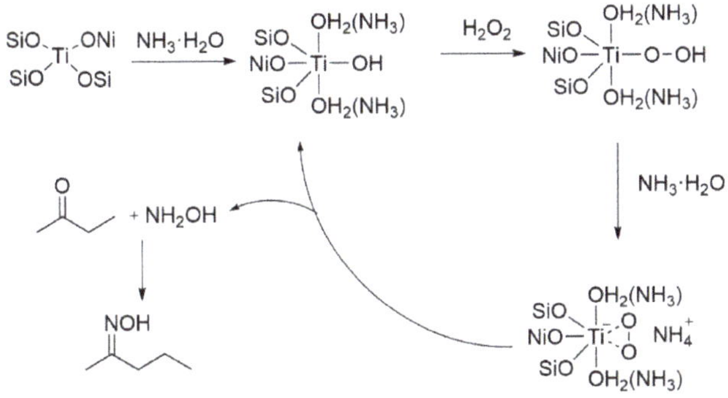

Scheme 2. Reaction mechanism of ammoximation of MEK catalyzed by Ni-TS-1.

3. Experimental Section

3.1. Materials

The industrial catalyst TS-1 (SiO_2/TiO_2 = 56, specific surface area of 394.7 m^2/g) was provided by Luxi Chemical Group Co., Ltd. Analytical grade ammonia (25%, AR, Laiyang, Shandong, China), hydrogen peroxide (30%, AR, Laiyang, Shandong, China), t-butanol (≥99%, AR, Shanghai, China), MEK (≥99%, AR, local vendor), nickel nitrate hexahydrate (98%, AR, Shanghai, China), Cobalt nitrate hexahydrate (98%, AR, Shanghai, China), Ferric nitrate nine-hydrate (≥99%, AR, Tianjin, China), Copper nitrate trihydrate (≥99%, AR, Tianjin, China), Cerium nitrate hexahydrate (98%, AR, Shanghai, China), and Anhydrous sodium carbonate (≥99.8%, AR, Xuzhou NO.2 Reagent Factory) were all obtained commercially.

3.2. Methods

3.2.1. Catalyst Preparation

The transition metal-modified TS-1 was prepared by an ultrasonic impregnation method [20]. A total of 4.00 g TS-1 was dispersed in 20 mL of deionized water, and sodium carbonate (Na_2CO_3:M^{n+} = 2:1) was added as precipitator. Under vigorous stirring, the corresponding aqueous solution of metal nitrate with certain quality was slowly dripped. The mixture was sonicated for 1 h and stirred vigorously for 3 h. Then the mixture was filtered and the solid was dried at 120 °C overnight and calcined at 550 °C for 6 h to obtain M-TS-1 (M is Fe, Co, Ni, Cu, Ce). The loading X is defined as m (metal element)/m (TS-1), and the range of X is 1–5 wt %.

3.2.2. Catalyst Characterization

Powder X-ray diffraction (XRD) was performed on a Rigaku Smartlab SE X-ray diffractometer with a Cu-kα radiation source at a tube voltage of 40 kV and a tube current of 25 mA. Diffraction patterns were recorded between 5° and 60°. FI-IR spectroscopy was measured on a Nicolet 380 spectrometer over a range of 4000 cm^{-1}–400 cm^{-1} at a resolution of 4 cm^{-1}. N_2 adsorption–desorption isotherms were collected by a Gemini VII analyzer. Prior to the tests, about 100 mg samples were degassed at 250 °C for 4 hours. Then, the sample completing the pretreatment was subjected to a nitrogen adsorption–desorption isotherm test at a liquid nitrogen temperature (77 K). DR UV-Vis was performed on a TU-1901 spectrophotometer in the wavelength range from 800 to 190 nm. The induced coupled plasma-atomic emission spectroscopy (ICP-AES) was carried out using a 1000-type spectroscopy. The sample of the catalyst is completely dissolved by 40% HF, and the sample was made up to 50 mL with water after HF removal by electric heating plate. The metal content in the catalyst is quantitatively detected. The loading efficiency (e) of Ni on TS-1 was calculated based on the actual loading of Ni element measured and the amount of Ni in the impregnating solution. TGA measurements were rendered on a DTG60A thermogravimetric analyzer from room temperature to 800 °C at the rate of 10 °C/min. The weight loss rate was the ratio of weight loss mass of catalyst to original mass of catalyst. The X-ray photoelectron spectroscopy was collected by A ESCALAB250 spectroscopy. Scanning electron microscopy (SEM) was performed on a JSM-6700F cold field high-resolution emission scanning electron microscope.

3.2.3. Point of Zero Charge (PZC) of Catalyst

The point of zero charge of catalyst (pH_{PZC}) was determined by the salt titration method: 0.4 mol/L $NaNO_3$ solution was prepared. We took 20 mL $NaNO_3$ solution and adjusted the pH to 2, 3, 4, 5, 6, 7, and 8 with NaOH or H_2SO_4. Then, 0.2 g catalyst was added to shake for 0.5 h until the pH was stabilized, and the pH change value (ΔpH) before and after the addition of catalyst was calculated. The initial pH was horizontal coordinate and ΔpH was vertical coordinate. When the ordinate is zero, the pH value is PZC.

3.3. Catalytic Activity Test

A total of 0.1 mol MEK, 1.00 g catalyst, and 25 mL t-butanol were placed in a four-neck round bottom flask with a condensation reflux device. The constant temperature water bath was used to keep the temperature of the reaction system at 343K. $NH_3 \cdot H_2O$ and H_2O_2 were slowly dropped into the flask for 1.5 h ($MEK:NH_3 \cdot H_2O:H_2O_2$ = 1:4:1.1), and then the reaction lasted for a certain period of time after the end of dropping. After the reaction, the catalysts were filtrated and separated. The content of MEK (standard curve method) was detected by high performance liquid chromatography (HPLC). The content of MEKO (standard curve method) was detected by gas chromatography (GC) after the filtrate was volumed with absolute ethanol.

The 3 wt % Ni-TS-1 catalyst was reused. The catalyst, for the first reaction, was completely transferred to the second reaction, and so on. The catalyst was reused for five times with other conditions unchanged. In addition, the catalyst that was reused five times was calcined (540 °C for 7 h) to activate and regenerate, and applied to the ammoximation of MEK again.

$$\text{Conversion of MEK (\%)} = \frac{\text{Converted moles of MEK}}{\text{Initial moles of MEK}} \times 100\%, \qquad (1)$$

$$\text{Selectivity of MEKO (\%)} = \frac{\text{Moles of MEKO}}{\text{Converted moles of MEK}} \times 100\%. \qquad (2)$$

Liquid chromatography conditions: Shimadzu HPLC LC-20AT; the column is Diamons C18 column (250 mm × 4.6 mm, 5 μm), the UV detection wavelength is 274 nm, the mobile phase is V $_{methanol}$:V $_{water}$ = 50:50, and the flow rate is 0.8 mL/min.

Gas chromatographic conditions: Shimadzu gas chromatograph GC-2010; the column is FFAP capillary column, the carrier gas is high purity nitrogen, the inlet and the detection port temperature is 200 °C, and the column temperature is 100 °C.

4. Conclusions

Five kinds of transition metal-modified TS-1 catalysts were prepared and employed to catalyze the ammoximation of MEK. The catalytic activity of various transition metal-modified TS-1 was diverse. Compared with TS-1 and other M-TS-1, 3 wt % Ni-TS-1 showed the best catalytic activity. The conversion of MEK and the selectivity of MEKO reached 99.0% and 99.3%, respectively. Furthermore, the catalyst was fully recovered without considerable decrease of MEK conversion and MEKO selectivity. The characterization of Ni-TS-1 showed that the introduction of nickel species did not change the skeleton structure of TS-1. Meanwhile, the nickel entering the framework of TS-1 effectively changed the electron environment of the Ti active sites and increased the electrophilicity of TS-1. In addition, the surface nickel species reduced the surface acidity of the catalyst, inhibiting the deep oxidation of the MEKO. We believe the obtained metal-modified TS-1 catalyst is very promising for the ammoximation of MEK and relative reactions for industrial applications.

Author Contributions: S.J. and X.X. conceived and designed the experiments and guided the research. D.Y., H.W., W.W., S.P., and X.Y. performed characterization and catalytic studies. All authors analyzed and discussed the results. D.Y., S.J., and X.X. analyzed the data and wrote the paper.

Funding: This research received no external funding.

Acknowledgments: This research was financially supported by the science and technology development plan of Shandong Province (2013GZX20109).

Conflicts of Interest: The authors declare no conflict of interest.

References

1. Hu, Y.; Dong, C.; Wang, T.; Luo, G. Cyclohexanone ammoximation over TS-1 catalyst without organic solvent in a microreaction system. *Chem. Eng. Sci.* **2018**, *187*, 60–66. [CrossRef]
2. Chu, Q.; Wang, P.; He, G.; Li, M.; Zhu, H.; Liu, R.; Pei, F. Continuous three-phase 2-butanone ammoximation process via spray forming TS-1 microspheres in a highly efficient jet loop reactor. *Chem. Eng. J.* **2017**, *325*, 169–175. [CrossRef]
3. Xue, X.; Song, F.; Ma, B.; Yu, Y.; Li, C.; Ding, Y. Selective ammoximation of ketones and aldehydes catalyzed by a trivanadium-substituted polyoxometalate with H_2O_2 and ammonia. *Catal. Commun.* **2013**, *33*, 61–65. [CrossRef]
4. Song, F.; Liu, Y.; Wang, L.; Zhang, H.; He, M.; Wu, P. Highly selective synthesis of methyl ethyl ketone oxime through ammoximation over Ti-MWW. *Appl. Catal. A Gen.* **2007**, *327*, 22–31. [CrossRef]
5. Jin, H.; Meng, C.; He, G.; Guo, X.; Yang, S. Green synthesis of acetaldehyde oxime using ammonia oxidation in the TS-1/H_2O_2 system. *React. Kinet. Mech. Catal.* **2018**, *125*, 1113–1125. [CrossRef]
6. Wang, W.; Fu, Y.; Guo, Y.; Guo, Y.; Gong, X.Q.; Lu, G. Preparation of lamellar-stacked TS-1 and its catalytic performance for the ammoximation of butanone with H_2O_2. *J. Mater. Sci.* **2017**, *53*, 4034–4045. [CrossRef]
7. Lu, T.; Zou, J.; Zhan, Y.; Yang, X.; Wen, Y.; Wang, X.; Zhou, L.; Xu, J. Highly Efficient Oxidation of Ethyl Lactate to Ethyl Pyruvate Catalyzed by TS-1 Under Mild Conditions. *ACS Catal.* **2018**, *8*, 1287–1296. [CrossRef]
8. Marco, T.; Giovanni, P.; Bruno, N. Preparation of Porous Crystalline Synthetic Material Comprised of Silicon and Titanium Oxides. U.S. Patent 4,410,501, 18 October 1983.
9. Pei, X.; Liu, X.; Liu, X.; Shan, J.; Fu, H.; Xie, Y.; Yan, X.; Meng, X.; Zheng, Y.; Li, G.; et al. Synthesis of Hierarchical Titanium Silicalite-1 Using a Carbon-Silica-Titania Composite from Xerogel Mild Carbonization. *Catalysts* **2019**, *9*, 672. [CrossRef]

10. Xue, Y.; Zuo, G.; Wen, Y.; Wei, H.; Liu, M.; Wang, X.; Li, B. Seed-assisted synthesis of TS-1 crystals containing Al with high catalytic performances in cyclohexanone ammoximation. *RSC Adv.* **2019**, *9*, 2386–2394. [CrossRef]
11. Xue, Y.; Wen, Y.; Wei, H.; Liu, M.; Huang, X.; Ye, X.; Wang, X.; Li, B. Hollow TS-1 mesocrystals: Hydrothermal construction and high catalytic performances in cyclohexanone ammoximation. *RSC Adv.* **2015**, *5*, 51563–51569. [CrossRef]
12. Zhuo, Z.; Wang, L.; Zhang, X.; Wu, L.; Liu, Y.; He, M. Insights into the key to highly selective synthesis of oxime via ammoximation over titanosilicates. *J. Catal.* **2015**, *329*, 107–118. [CrossRef]
13. Zhuo, Z.; Wu, L.; Wang, L.; Ding, Y.; Zhang, X.; Liu, Y.; He, M. Lewis acidic strength controlled highly selective synthesis of oxime via liquid-phase ammoximation over titanosilicates. *RSC Adv.* **2014**, *4*, 55685–55688. [CrossRef]
14. Wei, Y.; Li, G.; Su, R.; Lu, H.; Guo, H. Ti-sites environment-mediated hierarchical TS-1 catalyzing the solvent-free epoxidation: The remarkably promoting role of alcohol modification. *Appl. Catal. A Gen.* **2019**, *582*, 117108. [CrossRef]
15. Meng, C.; Yang, S.; He, G.; Luo, G.; Xu, X.; Jin, H. The Reaction Mechanism of Acetaldehyde Ammoximation to Its Oxime in the TS-1/H_2O_2 System. *Catalysts* **2016**, *6*, 109. [CrossRef]
16. Kumar, N.; Leino, E.; Mäki-Arvela, P.; Aho, A.; Käldström, M.; Tuominen, M.; Laukkanen, P.; Eränen, K.; Mikkola, J.P.; Salmi, T.; et al. Synthesis and characterization of solid base mesoporous and microporous catalysts: Influence of the support, structure and type of base metal. *Microporous Mesoporous Mater.* **2012**, *152*, 71–77. [CrossRef]
17. Fan, J.; Chen, A.; Chen, S.; Han, X.; Sun, H. Synthesis of Fe,Co,Ni loaded MCM-41 mesoporous molecular sieves and their catalytic oxidation performance. *Environ. Chem.* **2016**, *35*, 1116–1124.
18. Wu, G.; Xiao, J.; Zhang, L.; Wang, W.; Hong, Y.; Huang, H.; Jiang, Y.; Li, L.; Wang, C. Copper-modified TS-1 catalyzed hydroxylation of phenol with hydrogen peroxide as the oxidant. *RSC Adv.* **2016**, *6*, 101071–101078. [CrossRef]
19. Chen, S.; Ciotonea, C.; Ungureanu, A.; Dumitriu, E.; Catrinescu, C.; Wojcieszak, R.; Dumeignil, F.; Royer, S. Preparation of nickel (oxide) nanoparticles confined in the secondary pore network of mesoporous scaffolds using melt infiltration. *Catal. Today* **2019**, *334*, 48–58. [CrossRef]
20. Wu, M.; Chou, L.; Song, H. Effect of metals on titanium silicalite TS-1 for butadiene epoxidation. *Chin. J. Catal.* **2013**, *34*, 789–797. [CrossRef]
21. Wu, M.; Zhao, H.; Yang, J.; Zhao, J.; Song, H.; Chou, L. Precursor effect on the property and catalytic behavior of Fe-TS-1 in butadiene epoxidation. *Russ. J. Phys. Chem. A* **2017**, *91*, 2103–2109. [CrossRef]
22. Capel-Sanchez, M. Impregnation treatments of TS-1 catalysts and their relevance in alkene epoxidation with hydrogen peroxide. *Appl. Catal. A Gen.* **2003**, *246*, 69–77. [CrossRef]
23. Ma, Y.; Wu, Y.; Li, J.; Jiang, C. Oxidative degradation of nitrobenzene catalyzed by Cu^{2+}-H_2O_2 system in copper rinse water. *Chin. J. Environ. Eng.* **2016**, *10*, 4775–4782.
24. Tang, Q.; Zhang, Q.; Wu, H.; Wang, Y. Epoxidation of styrene with molecular oxygen catalyzed by cobalt(II)-containing molecular sieves. *J. Catal.* **2005**, *230*, 384–397. [CrossRef]
25. Vayssilov, G.N. Structural and Physicochemical Features of Titanium Silicalites. *Catal. Rev.* **1997**, *39*, 209–251. [CrossRef]
26. Wu, M.; Chou, L.; Song, H. Epoxidation of Butadiene Over Nickel Modified TS-1 Catalyst. *Catal. Lett.* **2012**, *142*, 627–636. [CrossRef]
27. Bordiga, S.; Damin, A.; Bonino, F.; Lamberti, C. Single Site Catalyst for Partial Oxidation Reaction: TS-1 Case Study. *Surf. Interfacial Organomet. Chem. Catal.* **2005**, *16*, 37–68.
28. Jorda, E.; Tuel, A.; Teissier, R.; Kervennal, J. TiF_4: An original and very interesting precursor to the synthesis of titanium containing silicalite-1. *Zeolites* **1997**, *19*, 238–245. [CrossRef]
29. Balducci, L.; Bianchi, D.; Bortolo, R.; D'Aloisio, R.; Ricci, M.; Tassinari, R.; Ungarelli, R. Direct oxidation of benzene to phenol with hydrogen peroxide over a modified titanium silicalite. *Angew. Chem.* **2003**, *42*, 4937–4940. [CrossRef]
30. Singh, H.; Rai, A.; Yadav, R.; Sinha, A.K. Glucose hydrogenation to sorbitol over unsupported mesoporous Ni/NiO catalyst. *Mol. Catal.* **2018**, *451*, 186–191. [CrossRef]
31. Li, J.; Qian, X.; Peng, Y.; Lin, J. Hierarchical structure NiO/CdS for highly performance H_2 evolution. *Mater. Lett.* **2018**, *224*, 82–85. [CrossRef]

32. Zhang, Z.; Kong, X.X.; Feng, M.; Luo, Z.H.; Lu, H.; Cao, G.P. In Situ Synthesis of TS-1 on Carbon Nanotube Decorated Nickel Foam with Ultrafine Nanoparticles and High Content of Skeleton Titanium. *Ind. Eng. Chem. Res.* **2018**, *58*, 69–78. [CrossRef]
33. Akhtar, M.S.; Alam, M.A.; Tauer, K.; Hossan, M.S.; Sharafat, M.K.; Rahman, M.M.; Minami, H.; Ahmad, H. Core-shell structured epoxide functional NiO/SiO$_2$ nanocomposite particles and photocatalytic decolorization of congo red aqueous solution. *Colloids Surf. A Physicochem. Eng. Asp.* **2017**, *529*, 783–792. [CrossRef]
34. Phanichphant, S.; Nakaruk, A.; Channei, D. Photocatalytic activity of the binary composite CeO$_2$/SiO$_2$ for degradation of dye. *Appl. Surf. Sci.* **2016**, *387*, 214–220. [CrossRef]
35. Mahmood, T.; Saddique, M.T.; Naeem, A.; Westerhoff, P.; Mustafa, S.; Alum, A. Comparison of Different Methods for the Point of Zero Charge Determination of NiO. *Ind. Eng. Chem. Res.* **2011**, *50*, 10017–10023. [CrossRef]
36. Zecchina, A.; Bordiga, S.; Lamberti, C.; Ricchiardi, G.; Scarano, D.; Petrini, G.; Leofanti, G.; Mantegazza, M. Structural characterization of Ti centres in Ti-silicalite and reaction mechanisms in cyclohexanone ammoximation. *Catal. Today* **1996**, *32*, 97–106. [CrossRef]
37. Chu, Q.; He, G.; Xi, Y.; Wang, P.; Yu, H.; Liu, R.; Zhu, H. Green synthesis of low-carbon chain nitroalkanes via a novel tandem reaction of ketones catalyzed by TS-1. *Catal. Commun.* **2018**, *108*, 46–50. [CrossRef]

© 2019 by the authors. Licensee MDPI, Basel, Switzerland. This article is an open access article distributed under the terms and conditions of the Creative Commons Attribution (CC BY) license (http://creativecommons.org/licenses/by/4.0/).

Communication

Carbozincation of Substituted 2-Alkynylamines, 1-Alkynylphosphines, 1-Alkynylphosphine Sulfides with Et₂Zn in the Presence of Catalytic System of Ti(O-*i*Pr)₄ and EtMgBr

Rita N. Kadikova *, Ilfir R. Ramazanov, Azat M. Gabdullin, Oleg S. Mozgovoi and Usein M. Dzhemilev

Institute of Petrochemistry and Catalysis of Russian Academy of Sciences, 141 Prospekt Oktyabrya, 450075 Ufa, Russia; ilfir.ramazanov@gmail.com (I.R.R.); saogabdullinsao@gmail.com (A.M.G.); skill15@mail.ru (O.S.M.); ink@anrb.ru (U.M.D.)
* Correspondence: kadikritan@gmail.com; Tel.: +7-(347)-284-27-50

Received: 1 November 2019; Accepted: 28 November 2019; Published: 4 December 2019

Abstract: The EtMgBr and Ti(O-*i*Pr)₄-catalyzed ethylzincation of 1-alkynylphosphine sulfides with Et₂Zn in diethyl ether followed by hydrolysis and deuterolysis affords corresponding β,β-disubstituted 1-alkenylphosphine sulfides with high yield. The EtMgBr and Ti(O-*i*Pr)₄-catalyzed reaction of 2-alkynylamines, 1-alkynylphosphines, and 1-alkenylphosphine sulfides with Et₂Zn in various solvents was studied. It has been found that the reaction of 2-alkynylamines and 1-alkynylphosphines in methylene chloride, hexane, toluene, benzene, and anisole leads to the selective formation of 2-alkenylamines and 1-alkenylphosphine oxides after oxidation with H_2O_2.

Keywords: ethylzincation; 1-alkynylphosphine sulfides; 1-alkynylphosphines; 2-alkynylamines; 2-zincoethylzincation; titanium catalysis; diethylzinc

1. Introduction

Recently we found that the reaction of 1-alkynylphosphines and substituted propargylamines with Et₂Zn in the presence of catalytic amounts of reagents, such as titanium(IV) isopropoxide and ethylmagnesium bromide, gives products of triple bond 2-zincoethylzincation [1]. Our study supplements Negishi's work on Ti-Mg-catalyzed cyclozincation of α,ω-enynes with Et₂Zn well [2]. It is known that Cu-, Ni-, Co-, and Ru-catalyzed carbozincation of alkynyl sulfoximines, alkynyl sulfones [3], 1-alkynyl sulfoxides [4], propargylic ethers [5], propargyl alcohols [6], and ynamides [7,8] is an effective tool for the synthesis of polysubstituted functionalized olefins. As regards the carbozincation of inactivated alkynes, a limited number of examples are reported in the literature. Ni-catalyzed stereoselective carbozincation of but-1-yn-1-ylbenzene with Ph₂Zn is known [9], as well as Co-catalyzed allylzincation of aryl alkynes [10]. As to the effective carbozincation of dialkyl substituted alkynes, only ethylzincation [6] and allylzincation [11] of dec-5-yne in the presence of stoichiometric amounts of Cp₂ZrI₂ were described in the literature. It should be emphasized that the use of a catalytic system consisting of EtMgBr and Ti(O-*i*Pr)₄ for 2-zincethylzincation of α,ω-enynes was first reported by Negishi [2]. Before our study, no examples of carbozincation of alkynylphosphines were known. As regards carbozincation of N-containing alkynes, only one example of diastereoselective carbozincation of lithiated (R)-(1)-N-methyl-N-propargyl-1-phenylethylamine with crotylzinc bromide was reported [12]. Therefore, further investigation of the scope of applicability of Ti-Mg-catalyzed carbozincation of functionally substituted acetylene derivatives was of considerable interest. In view of the high demand for ligands based on phosphines as well as their derivatives in organometallic and coordination chemistry [13,14], we studied, first of all, the catalytic carbozincation of 1-alkynylphosphine oxides and -phosphine sulfides.

2. Results and Discussions

2.1. Ti-Mg-Catalyzed Carbozincation of Substituted 1-Alkynylphosphine Sulfides with Et_2Zn

Unfortunately, none of our attempts to perform carbozincation of substituted 1-alkynylphoshine oxides [diphenyl(phenylethynyl)phosphine oxide, hept-1-yn-1-yl(diphenyl)phosphine oxide] with 2.5 equivalents of Et_2Zn (1 M in hexane) in the presence of 0.15 equivalent of $Ti(O\text{-}iPr)_4$ (0.3 M in hexane) and 0.2 equivalent of EtMgBr (2.5 M in Et_2O) in diethyl ether solution at room temperature met with success. However, 1-alkynylphosphine sulfides (hex-1-yn-1-yldiphenylphosphine sulfide, hept-1-yn-1-yldiphenylphosphine sulfide, oct-1-yn-1-yldiphenylphosphine sulfide) proved to be reactive in this reaction. We found that 1-alkynylphosphine sulfides **1** react with 2.5 equivalents of Et_2Zn (1 M in hexane) in the presence of 0.2 equivalent of EtMgBr (2.5 M in Et_2O) and 0.15 equivalent of $Ti(O\text{-}iPr)_4$ (0.3 M in hexane) in diethyl ether at room temperature for 18 h to give, after hydrolysis or deuterolysis, the corresponding substituted Z-1-alkenylphosphine sulfides **3a–c** and **4a** in high yields (Figure 1).

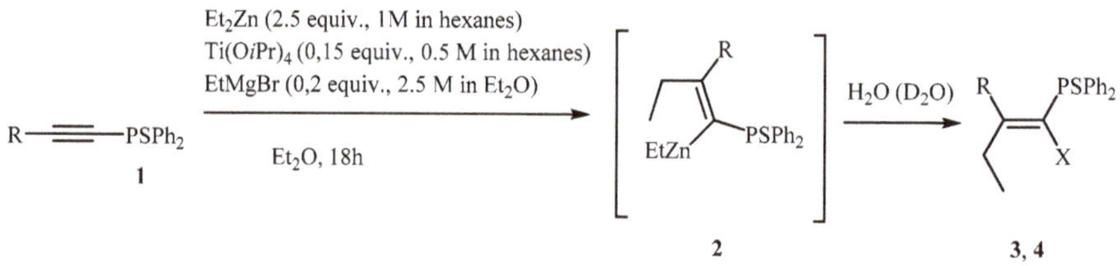

Figure 1. Ti-Mg-catalyzed carbozincation of substituted 1-alkynylphosphine sulfides with Et_2Zn in Et_2O.

The crucial difference from the reaction of 1-alkynylphosphines reported in our previous study [1] is that 1-alkynylphosphine sulfides are converted to ethylzincation rather than 2-zincoethylzincation products under the same conditions. In our opinion, the formation of ethylzincation products from alkynylphosphine sulfides proceeds in the following way. According to the presented Figure 2, fast ligand exchange between titanium(IV) isopropoxide and ethylmagnesium bromide yields an unstable diethyltitanium compound, which is then converted to a titanacyclopropane intermediate (titanium(II)–ethylene complex). Generation of a titanacyclopropane complex upon the reaction of Grignard reagents with titanium (IV) alkoxides was first suggested by Kulinkovich [15]. According to Figure 2, the subsequent insertion of 1-alkynylphosphine sulfide into the titanium-carbon bond of titanacyclopropane intermediate **A** results in the formation of titanacyclopentene intermediate **B**. The ligand exchange between intermediate **B** and the Et_2Zn molecule yields bimetallic intermediate **C**. The formation of a similar bimetallic complex is postulated in the Zr-catalyzed ethylmagnesiation of inactivated alkenes [16]. The subsequent hydrogen transfer from the ethyl group at the titanium atom of the bimetallic complex C regenerates the titanacyclopropane intermediate and affords ethylzincation product **D**.

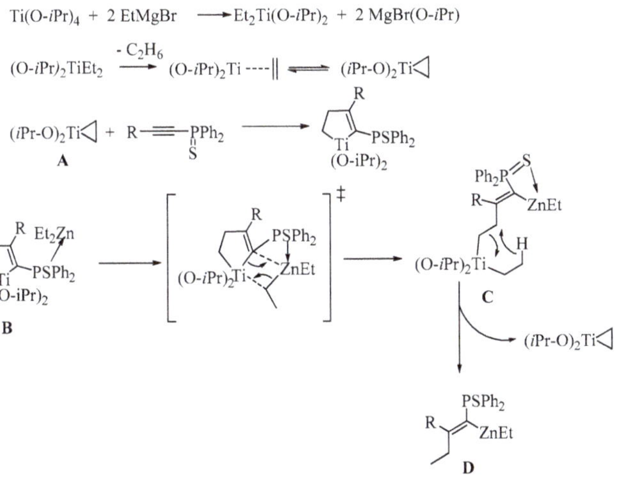

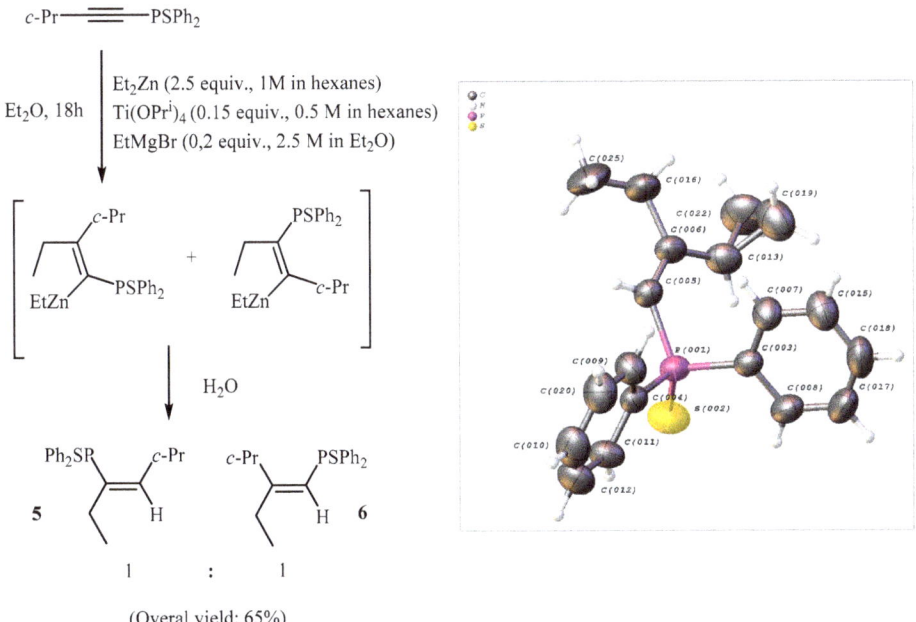

Figure 2. The plausible mechanism of Ti-Mg-catalyzed carbozincation of substituted 1-alkynylphosphine sulfides with Et_2Zn.

Unlike reactions of alkyl-substituted alkynylphosphine sulfides, the reaction of (cyclopropylethynyl) diphenylphosphine sulfide is not regioselective and gives a mixture of regioisomers of 1-alkenylphosphines **5** and **6** with Z-configuration of the double bond in 1:1 ratio (Figure 3).

Figure 3. Ti-Mg-catalyzed carbozincation of substituted (cyclopropylethynyl) diphenylphosphine sulfide with Et_2Zn.

The structure of regioisomer **6** with the geminal location of cyclopropyl and ethyl fragments at the double-bond carbon was defined by X-ray diffraction. Presumably, one of the factors responsible for the observed non-selective transformation of cyclopropyl-substituted alkynylphosphine sulfide is the presence of C–C agostic interaction between the titanium atom and cyclopropane ring. The agostic interaction involving cyclopropane moieties was reported for Pt complexes, such as [PtCl$_2$(c-C$_3$H$_6$)]$_2$ and PtCl$_2$(c-C$_3$H$_6$)(py)$_2$ (py = pyridine) [17], and for lithium cyclopropoxide complex [18,19].

Thus, depending on the substituent, Ti-Mg-catalyzed reaction of functionally substituted alkynes with Et$_2$Zn follows either a 2-zincoethylzincation (1-alkynylphosphines, 2-alkynylamines) [1] or ethylzincation (1-alkynylphosphine sulfides) pathway. In this respect, it was interesting to study the behavior of acetylenic alcohols and ethers in this reaction. Unfortunately, our attempts to perform this reaction with hept-2-yn-1-ol, oct-3-yn-1-ol, or (hept-2-yn-1-yloxy) benzene in diethyl ether failed. Probably, coordination of the titanium ethylene complex (which can be represented as an equivalent of divalent Ti(O-iPr)$_2$ stabilized by ethylene ligand [20]) to the oxygen atom of phosphine oxide, alcohol, or ether group gives a stable unreactive organometallic complex. This complex formation inhibits titanium coordination to the triple carbon–carbon bond and thus prevents the formation of titanacyclopentene.

2.2. The EtMgBr and Ti(O-iPr)$_4$-Catalyzed 2-Zincoethylzincation of Substituted Substituted Propargylamines with Et$_2$Zn

In view of the presumed importance of various ligand coordination effects for this reaction, we studied the EtMgBr and Ti(O-iPr)$_4$-catalyzed reaction of substituted propargylamines with Et$_2$Zn in solvents of different nature. The reaction of N,N-dimethylbut-2-ynyl-1-amine **7a** with 2.5 equivalents of Et$_2$Zn (1 M in hexane) in the presence of 0.15 equivalent of Ti(O-iPr)$_4$ (0.3 M in hexane) and 0.2 equivalent of EtMgBr (2.5 M in Et$_2$O) was equally efficient in diethyl ether, anisole, dichloromethane, hexane, benzene, or toluene and resulted in regio- and stereoselective formation of dideuterated allylamine **8a** with Z-configuration (Figure 4). Similar results were obtained for N,N-dimethylundec-2-ynyl-1-amine, 1-(hept-2-yn-1-yl)piperidine, N,N-dimethylnon-2-ynyl-1-amine, and 4-(hept-2-yn-1-yl)morpholine. More evidence for the structure of the resulting allylamines was gained by converting them to iodinolysis products **10c** and **10e**. Meanwhile, N,N-dimethylbut-2-ynyl-1-amine **7a** proved to be completely inert towards the reaction carried out in 1,4-dioxane, tetrahydrofuran, 1,2-dichloroethane, 1,2-dimethoxyethane, chloroform, or triethylamine.

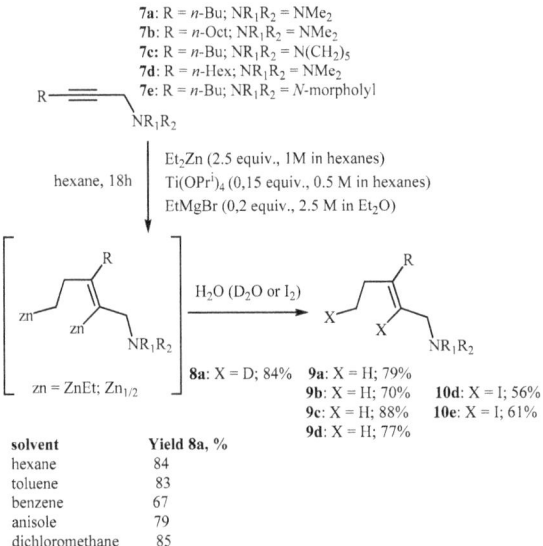

Figure 4. EtMgBr and Ti(O-*i*Pr)$_4$-catalyzed 2-zincoethylzincation of substituted propargylamines with diethylzinc in various solvents.

We suggested that in the case of 1,2-dimethoxyethane, 1,4-dioxane, tetrahydrofuran, and triethylamine, the acetylenic substrate molecule cannot displace the solvent molecule from the coordination sphere of the low-valence titanium atom in intermediate **E** (Figure 5) and, hence, intermediate **F** is not formed and the catalytic cycle is interrupted. Quantum chemical B3LYP/6-31G(d,p) modeling of the step of displacement of a solvent molecule by *N,N*-dimethylbut-2-ynyl-1-amine, which was chosen as the model compound, demonstrated that the ease of displacement (Gibbs free energy) increases in the series Et$_3$N (−3.1 kcal/mol) < THF (−4.9 kcal/mol) < Me$_2$O (−6.5 kcal/mol). According to quantum chemical calculations, for dichloromethane, hexane, or aromatic hydrocarbons (benzene, toluene) as solvents, the equilibrium between intermediates **A** and **E** is shifted towards the non-solvated titanacyclopropane **A**, which facilitates the formation of intermediate **F**.

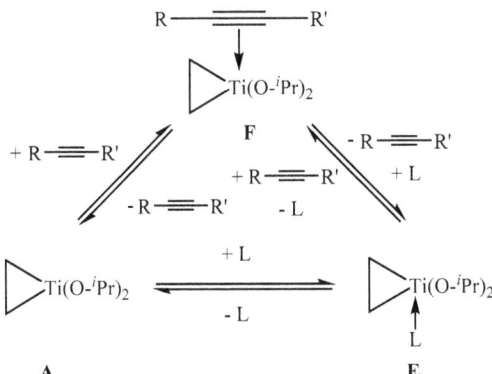

Figure 5. Ligand exchange in the coordination sphere of the titanium atom of titanacyclopropane intermediate.

Despite similar natures of dichloromethane, 1,2-dichloroethane, and chloroform, the reaction smoothly proceeds in dichloromethane, but does not take place in 1,2-dichloroethane or chloroform.

In our opinion, this difference may be attributable to the instability of chloroform and 1,2-dichloroethane under conditions of reaction with EtMgBr and Ti(O-iPr)$_4$. The use of these solvents in organomagnesium chemistry is fairly limited. For instance, it is known that phenylmagnesium bromide and ethymagnesium iodide readily react with chloroform and tetrachloromethane to give dihalocarbenes [21]. On the other hand, there are many examples of cross-coupling reactions of Grignard reagents with polychlorinated solvents activated by transition metal catalysts [22–26].

2.3. EtMgBr and Ti(O-iPr)$_4$-Catalyzed 2-Zincoethylzincation of Substituted 1-Alkynylphosphines with Diethylzinc

In connection with the obtained results, we were interested in studying the effect of various solvents on the EtMgBr and Ti(O-iPr)$_4$-catalyzed reaction of P-containing alkynes—1-alkynylphosphines, 1-alkynylphosphine sulfides, and 1-alkynylphosphine oxides with Et$_2$Zn. The reaction of substituted 1-alkynylphosphines **11** with 2.5 equivalents Et$_2$Zn (1 M in hexanes) in the presence of 0.15 equivalent Ti(O-iPr)$_4$ (0.3 M in hexanes) and 0.2 equivalent EtMgBr (2.5 M in Et$_2$O) at room temperature followed by oxidation with an aqueous solution of H$_2$O$_2$ (37%) or sulfuration with elemental sulfur is equally effective in diethyl ether [1], methylene chloride, hexane, and toluene with regio- and stereoselective formation of the corresponding 1-alkenylphosphine oxides and sulfides of the Z-configuration **12a**, **13b,c** and **14c** (Figure 6).

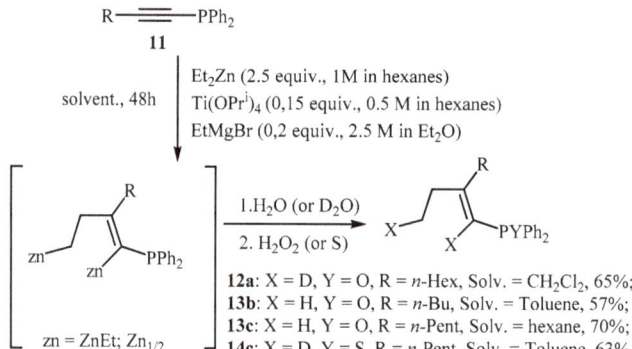

Figure 6. Ti-Mg-catalyzed carbozincation of substituted 1-alkynylphosphines with Et$_2$Zn in various solvents.

It should be noted that for the complete conversion of 1-alkynylphosphines **11** at room temperature in methylene chloride, toluene, and hexane, about 48 h are required. An increase in temperature to 40 °C leads to a deterioration in the selectivity of the reaction and the formation of difficult-to-analyzemixture of products. As expected, hept-1-yn-1-yldiphenylphosphine oxide was inert not only in diethyl ether (as described above) but also in methylene chloride, toluene, and hexane. At the same time, the Ti-Mg-catalyzed reaction of substituted 1-alkynylphosphine sulfides with Et$_2$Zn in methylene chloride, toluene, and hexane does not proceed stereoselectively and leads to the formation of a mixture of stereoisomers. For example, the reaction of hept-1-yn-1-yldiphenylphosphine sulfide **15** with 2.5 equiv. Et$_2$Zn (1 M in hexanes) in the presence of 15 mol. % Ti (O-iPr)$_4$ (0.3 M in hexanes) and 20 mol. % EtMgBr (2.5 M in Et$_2$O) in methylene chloride leads to the formation of a mixture of **16** (Z)- and **17** (E)-isomers in a 2:1 ratio with a total yield of 71% (Figure 7).

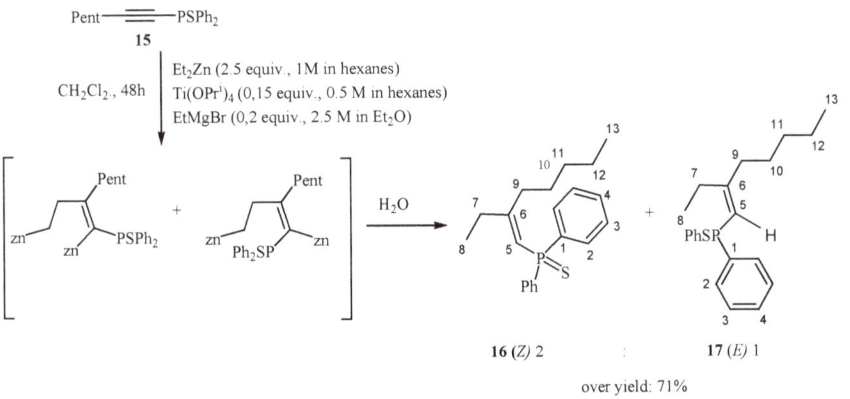

Figure 7. Ti-Mg-catalyzed carbozincation of substituted hept-1-yn-1-yldiphenylphosphine sulfide.

The formation of an isomeric mixture is indicated in the ^{13}C NMR spectrum of the reaction products by the presence of a double set of signals in a 2:1 ratio of the following carbon atoms: C-6 (δ 168.3 ppm and δ 168.1 ppm), C-7 (δ 31.2 ppm and δ 27.1 ppm), C-8 (δ 12.2 ppm and δ 11.5 ppm), C-9 (δ 34.2 ppm and δ 37.9 ppm), C-10 (δ 27.1 ppm and δ 27.5 ppm), C-11 (δ 31.9 ppm and δ 31.6 ppm)), C-12 (δ 22.4 ppm and δ 22.5 ppm), C-13 (δ 13.9 ppm and δ 14.1 ppm). The Overhauser effects observed in the NOESY spectrum between the methylene group H$_2$C-9 (δ 2.39 ppm) and the protons of the aromatic substituent of the compound **16**, as well as the cross-interaction between the protons H$_2$C-7 (δ 2.27 ppm) and HC-5 (δ 6.03 ppm) of the compound **17** allowed us to identify the obtained adducts as Z- and E-isomers, respectively.

3. Materials and Methods

3.1. Materials

The reagents were obtained from Acros and Sigma-Aldrich. Hexane and dichloromethane were dried over P_2O_5. 1,4-Dioxane, diethyl ether, tetrahydrofuran, toluene, benzene, and anisole were dried over sodium. Dried 1,2-dimethoxyethane was obtained from Sigma-Aldrich. 1-Alkynyl derivatives of phosphine oxides and 1-alkynyl phosphine sulfides **1** were prepared by the oxidation of 1-alkynyl phosphines with 30% aq. H_2O_2 [27] and based on the reaction of 1-alkynyl phosphines with sulfur [28], respectively. 1-Alkynylamines **7a**, **7b**, and **7d** were synthesized by aminomethylation of terminal alkynes by bisamine [29]. Alkynylamines **7c** and **7e** were prepared by aminomethylation of terminal alkynes with aqueous formaldehyde, 37 wt. % in H$_2$O and secondary amines using a CuI catalysis [30]. Nuclear magnetic resonance spectroscopy was performed. NMR spectra were recorded on a Brucker Avance 400 spectrometer at 400 MHz for ^{1}H and at 100 MHz for ^{13}C in CDCl$_3$. The numbering of atoms in the ^{1}H and ^{13}C NMR spectra of the compounds **3a–c, 4a, 5, 6, 8a, 9a–d, 10d, 10e, 12a, 13b,c, 14c, 16, 17** is shown in the Supplementary Materials. X-ray diffraction analysis was performed with an XCaliburEos diffractometer (graphite monochromator, MoKα radiation, λ = 0.71073 Å, w-scan mode, 2θmax = 62°). The data were treated using the CrysAlisProOxfordDiffractionLtd. program package, version 1.171.36.20. The refinement was done with the SHELX97 program package [31]. Elemental analysis was implemented with a Carlo-Erba CHN 1106 elemental analyzer. Mass spectra were obtained using a Finnigan 4021 instrument. The yields of chemical reaction products were obtained from the isolated amount of 2-alkenyl amines, 1-alkenyl phosphine oxides, and 1-alkenyl phosphine sulfides obtained from starting alkynes. All quantumchemical calculations were carried out using the B3LYP/6-31G(d,p) basis set (Gaussian 09 software) [32]. The ^{13}C NMR and ^{1}H NMR data of the products are shown in the Supplementary Materials.

3.2. Methods

3.2.1. Preparation of 1-Alkenyl Phosphine Sulfides 3a–c, 4a and 5, 6 via Ethylmagnesium Bromide and Titanium(IV) Isopropoxide-Catalyzed Reaction of 1-Alkynyl Phosphine Sulfides with Et$_2$Zn

(Z)-(2-Ethylhex-1-en-1-yl)diphenylphosphine sulfide (**3a**). Typical Procedure. To a solution of hex-1-yn-1-yldiphenylphosphine sulfide (596 mg, 2 mmol) and Et$_2$Zn (1 M in hexanes, 5 mL, 5 mmol) in ether (6 mL) was added Ti(O-iPr)$_4$ (0.5 M in hexanes, 0.6 mL, 0.3 mmol). Ethylmagnesium bromide (2.5 M in Et$_2$O, 0.16 mL, 0.4 mmol) was then was added dropwise, and the reaction mixture turned black. After 18 h at room temperature, the mixture was diluted with Et$_2$O (5 mL), and 25 wt % aq. KOH (3 mL) was added dropwise while the flask was cooled in an ice bath. The aqueous phase was extracted with Et$_2$O (3 × 5 mL). The combined organic phase was washed with a saturated aqueous solution of NaCl (10 mL) and dried over anhydrous MgSO$_4$. The reaction mixture was filtered and concentrated in vacuo to give the crude product as a yellow oil. Evaporation of the solvent and purification of the residue by column chromatography (hexane:ethyl acetate:methanol = 5:2:1)) gave **3a** (538 mg, 82%) as a colorless oil. R_f 0.42. ^{1}H NMR (δ, ppm, J/Hz): 0.73 (t, J = 7.3, 3H, C(12)H$_3$), 1.00–1.10 (m, 2H, C(11)H$_2$), 1.13 (t, J = 7.4, 2H, C(8)H$_3$), 1.20–1.30 (m, 2H, C(10)H$_2$), 2.29 (q, J = 7.5, 2H, C(7)H$_2$), 2.37 (t, J = 7.4, 2H, C(9)H$_2$), 6.03 (d, J = 23.5, 2H, C(7)H$_2$), 7.25–8.00 (m, 10H, Ph). ^{13}C NMR (δ, ppm, J/Hz): 12.17 (C(8)), 13.79 (C(12)), 22.80 (C(11)), 29.48 (C(10)), 31.21 (d, J = 16.4, C(7)), 33.92 (d, J = 9.3, C(9)), 115.99 (d, J = 89.4, C(5)), 128.43 (d, J = 12.3, 4C, C(3)), 131.01 (d, J = 2.5, 2C, C(4)), 131.20 (d, J = 10.5, 4C, C(2)), 135.21 (d, J = 84.2, 2C, C(1)), 168.22 (C(6)). ^{31}P NMR (δ, ppm): 28.68. MS (EI): m/z, % = 328 (45) [M$^+$], 299 (18), 254 (4), 218 (100), 183 (48), 139 (30), 108 (18), 41 (17). Anal. calcd for C$_{20}$H$_{25}$PS, (%): C, 73.14; H, 7.67. Found, %: C, 73.20; H, 7.71. The ^{1}H NMR and ^{13}C NMR of the compounds **3b,c**, **4a**, **5**, **6** data of coupling products were shown in the Supplementary Materials.

3.2.2. Preparation of 1-Alkenyl Phosphine Oxides 12a, 13b, 13c, 14c, 16, and 17 via Titanium(IV) Isopropoxide and Ethylmagnesium Bromide-Catalyzed Reaction of 1-Alkynyl Phosphines with Et$_2$Zn

(Z)-(2-(Ethyl-2-d)oct-1-en-1-yl-1-d)diphenylphosphine oxide (**12a**). Typical Procedure. To a solution of oct-1-yn-1-yldiphenylphosphane (588 mg, 2 mmol) and Et$_2$Zn (1 M in hexanes, 5 mL, 5 mmol) in dichloromethane (5 mL) was added Ti(OPr-i)$_4$ (0.5 M in hexanes, 0.6 mL, 0.3 mmol). Ether solution of EtMgBr (2.5 M in Et$_2$O, 0.16 mL, 0.4 mmol) was then added, and the mixture turned black. After 48 h at room temperature, CH$_2$Cl$_2$ (5 mL) was added to the reaction mixture, and D$_2$O (3 mL) was added dropwise while the flask was cooled in an ice bath. The aqueous inorganic layer was extracted through CH$_2$Cl$_2$ (3 × 5 mL). The combined organic phase was washed sequentially with water and brine (10 mL) and dried over anhydrous MgSO$_4$. The solvent was evaporated under reduced pressure. A 30% hydrogen peroxide solution (0.35 mL, 3 mmol) was slowly added dropwise with vigorous stirring to a solution of the crude residue (2-(Ethyl-2-d)oct-1-en-1-yl-1-d)diphenylphosphine oxide, in chloroform (5 mL). The reaction mixture was stirred for 1 h and washed with water (3 × 5 mL), the organic layer was dried over MgSO$_4$. The reaction mixture was filtered through a filter paper and concentrated in vacuo to give crude product as a yellow oil that was purified by column chromatography (silica gel, hexane:ethyl acetate:methanol = 5:2:1) to afford **12a** (445 mg, 65%). R_f 0.59. The spectral properties (^{1}H NMR, ^{13}C NMR, MS) were in good agreement with those that were reported in the literature [33]. The ^{1}H NMR and ^{13}C NMR of the compounds **13b**, **13c**, **14c**, **16**, **17** data of coupling products are shown in the Supplementary Materials.

3.2.3. Preparation of Allylic Amines 8a, 9a–d via Titanium(IV) Isopropoxide and Ethylmagnesium Bromide-Catalyzed Reaction of 2-Alkynylamines with Et$_2$Zn

(Z)-3-(Ethyl-2-d)-N,N-dimethylhept-2-en-1-amine-2-d (**8a**). Typical Procedure. To 278 mg of N,N-dimethylhept-2-yn-1-amine (2 mmol) and 5 mL of Et$_2$Zn (1 M in hexanes, 5 mmol) suspended in 6 ml hexane was added under an atmosphere of argon Ti(OPr-i)$_4$ (0.5 M in hexanes, 0.6 mL, 0.3 mmol). Then sequentially, EtMgBr (2.5 M in Et$_2$O, 0.16 mL, 0.4 mmol) was added, and the mixture turned black. After 18 h at room temperature, the reaction mixture was diluted with Et$_2$O (5 mL), and D$_2$O (3 mL) was added dropwise while the flask was cooled in an ice bath. The aqueous layer was extracted with Et$_2$O (3 × 5 mL). The organic layers were washed with brine (10 mL), dried over anhydrous CaCl$_2$. The reaction mixture was filtered through a filter paper and concentrated in vacuo to give crude product as a yellow oil. The residue was distilled through a micro column at 10 mmHg to give 8a (287 mg, 84%) as a colorless oil. b.p. 87–89 °C (10 mmHg) (lit. [7] b.p. 91–93 °C (15 mmHg)). ^{1}H NMR (400MHz, CDCl$_3$): δ = 0.92 (t, J = 6.3 Hz, 3H, C(11)H$_3$), 1.00 (t, J = 7.7 Hz, 3H, C(5)H$_3$), 1.25–1.40 (m, 4H, C(9,10)H$_2$), 2.03 (t, J = 7.8 Hz, 2H, C(4)H$_2$), 2.10–2.35 (m, 2H, C(8)H$_2$), 2.23 (s, 6H, C(6,7)H$_3$), 2.90 (s, 2H, C(1)H$_2$). ^{13}C NMR (100 MHz, CDCl$_3$): δ = 12.41 (t, C(5), $^1J_{CD}$ = 19.3 Hz), 14.02 (C(11)), 22.84 (C(10)), 29.41 and 30.30 and 30.71 (C(4,8,9)), 45.26 (2C(6,7)), 56.77 (C(1)), 144.27 (C(3)).

MS (EI): m/z, % = 171 (14) [M$^+$], 142 (10), 126 (18), 112 (21), 95 (100), 82 (32), 58 (49), 46 (48). Anal. calcd for C$_{11}$H$_{21}$D$_2$N, (%): C, 77.12. Found, %: C, 77.21. The ^{1}H NMR and ^{13}C NMR of the compounds 9a–d data of coupling products are shown in the Supplementary Materials.

3.2.4. The Iodination of Intermediate Organozinc Compounds

(Z)-2-Iodo-3-(2-iodoethyl)-N,N-dimethylnon-2-en-1-amine (10d). Typical Procedure. To a solution of N,N-dimethylnon-2-yn-1-amine (334 g, 2 mmol) and Et$_2$Zn (1 M in hexanes, 5 mL, 5 mmol) in toluene (6 mL) was added Ti(OPr-i)$_4$ (0.5 M in hexanes, 0.3 mL, 0.2 mmol) followed by ethylmagnesium bromide (2.5 M in Et$_2$O, 0.16 mL, 0.4 mmol). After 18 h at 23 C, the reaction mixture was cooled to −78 °C, and a solution of I$_2$ (1575 mg, 12.5 mmol) in THF (12.5 mL) was added via cannula. The mixture was warmed to room temperature and stirred for 10 h. The mixture was then treated by a 25% water solution of KOH and Et$_2$O. The organic phase was washed with water and an aqueous solution of Na$_2$S$_2$O$_3$, drying with MgSO$_4$. Evaporation of the solvent and purification of the residue by column chromatography (hexane/ethyl acetate, 5:1) gave a yellow oil; yield: 503 mg, (56%); R_f = 0.73 (hexane/ethyl acetate, 5:1). The spectral properties (^{1}H NMR, ^{13}C NMR, MS) of the compounds 10d, 10e were in good agreement with those that were reported in the literature [1].

4. Conclusions

Thus, Ti-Mg-catalyzed carbozincation of 1-alkynylphosphine sulfides with Et$_2$Zn in a solution of diethyl ether proceeds in a stereoselective manner, while the use of methylene chloride, hexane, and toluene as solvents leads to the formation of a mixture of stereoisomers. In the present work, it was also demonstrated that the selective Ti-Mg-catalyzed 2-zincethylzincation of 2-alkynylamines and 1-alkynylphosphines is possible not only in diethyl ether, as we showed earlier [1], but also in such solvents as hexane, methylene, benzene, toluene, and anisole. This study opens up further prospects for the use of metal complex catalyzed organozinc synthesis to create new methods for the production of olefins based on various transformations of functionalized acetylene derivatives.

Supplementary Materials: The following are available online at http://www.mdpi.com/2073-4344/9/12/1022/s1, analytical data and NMR spectrum for all compounds.

Author Contributions: Conceptualization and methodology, A.M.G. and O.S.M.; investigation and writing—original draft preparation, R.N.K.; writing—review and editing, I.R.R. and U.M.D.

Funding: This work was financially supported by the Russian Science Foundation (grant No. 19-73-10113).

Acknowledgments: We acknowledge the Center of collective use of the unique equipment «Agidel» at the Institute of Petrochemistry and Catalysis of the Russian Academy of Sciences.

Conflicts of Interest: The authors declare no conflict of interest.

References

1. Kadikova, R.N.; Ramazanov, I.R.; Mozgovoi, O.S.; Gabdullin, A.M.; Dzhemilev, U.M. 2-Zincoethylzincation of 2-Alkynylamines and 1-Alkynylphosphines Catalyzed by Titanium(IV) Isopropoxide and Ethylmagnesium Bromide. *Synlett* **2019**, *30*, 311–314. [CrossRef]
2. Montchamp, J.-L.; Negishi, E.-I. Carbozincation of Enynes Catalyzed by Titanium(IV) Alkoxides and Alkylmagnesium Derivatives. *J. Am. Chem. Soc.* **1998**, *120*, 5345–5346. [CrossRef]
3. Sklute, G.; Bolm, C.; Marek, I. Regio- and stereoselective copper-catalyzed carbozincation reactions of alkynyl sulfoximines and sulfones. *Org. Lett.* **2007**, *9*, 1259–1261. [CrossRef]
4. Maezaki, N.; Sawamoto, H.; Yoshigami, R.; Suzuki, T.; Tanaka, T. Geometrically Selective Synthesis of Functionalized β,β-Disubstituted Vinylic Sulfoxides by Cu-Catalyzed Conjugate Addition of Organozinc Reagents to 1-Alkynyl Sulfoxides. *Org. Lett.* **2003**, *5*, 1345–1347. [CrossRef]
5. Stüdemann, T.; Ibrahim-Ouali, M.; Knochel, P. A nickel-catalyzed carbozincation of aryl-substituted alkynes. *Tetrahedron* **1998**, *54*, 1299–1316. [CrossRef]
6. Yasui, H.; Nishikawa, T.; Yorimitsu, H.; Oshima, K. Bull. Cobalt-Catalyzed Allylzincations of Internal Alkynes. *Chem. Soc. Jpn.* **2006**, *79*, 1271–1274. [CrossRef]
7. Gourdet, B.; Lam, H.W. Stereoselective Synthesis of Multisubstituted Enamides via Rhodium-Catalyzed Carbozincation of Ynamides. *J. Am. Chem. Soc.* **2009**, *131*, 3802–3803. [CrossRef]
8. Sallio, R.; Corpet, M.; Habert, L.; Durandetti, M.; Gosmini, C.; Gillaizeau, I. Cobalt-Catalyzed Carbozincation of Ynamides. *J. Org. Chem.* **2017**, *82*, 1254–1259. [CrossRef]
9. Stüdemann, T.; Knochel, P. New Nickel-Catalyzed Carbozincation of Alkynes: A Short Synthesis of (2)-Tamoxifen. *Angew. Chem. Int. Ed.* **1997**, *36*, 93–95. [CrossRef]
10. Negishi, E.; Van Horn, D.E.; Yoshida, T.; Rand, C.L. Selective carbon-carbon bond formation via transition metal catalysis. 31. Controlled Carbometallation. 15. Zirconium-Promoted Carbozincation of Alkynes. *Organometallics* **1983**, *2*, 563–565. [CrossRef]
11. Negishi, E.; Miller, J.A. Selective carbon-carbon bond formation via transition metal catalysis. 37. Controlled carbometalation. 16. Novel syntheses of .alpha.,.beta.-unsaturated cyclopentenones via allylzincation of alkynes. *J. Am. Chem. Soc.* **1983**, *105*, 6761–6763. [CrossRef]
12. Rezaei, H.; Marek, I.; Normant, J.F. Diastereoselective carbozincation of propargylic amines. *Tetrahedron* **2001**, *57*, 2477–2483. [CrossRef]
13. Bertrand, G. Phosphorus chemistry: Introduction. *Chem. Rev.* **1994**, *94*, 1161–1162. [CrossRef]
14. Kollár, L.; Keglevich, G.P. Heterocycles as Ligands in Homogeneous Catalytic Reactions. *Chem. Rev.* **2010**, *110*, 4257–4302. [CrossRef]
15. Kulinkovich, O.G.; Sviridov, S.V.; Vasilevskii, D.A.; Pritytskaya, T.S. Reaction of ethylmagnesium bromide with carboxylic esters in the presence of tetraisopropoxytitanium. *Zh. Org. Khim.* **1989**, *25*, 2244–2245.
16. Lewis, D.P.; Whitby, R.J. The mechanism of the zirconium catalysed ethyl- and 2-magnesioethyl-magnesiation of unactivated alkenes. *Tetrahedron* **1995**, *51*, 4541–4550. [CrossRef]
17. Tipper, C.F.H. Some Reactions of cycloPropane, and a Comparison with the Lower Olefins. Part II.* Some Platinous-cycloPropane Complexes. *J. Chem. Soc.* **1955**, 2045–2046, Reprint Order No. 6013.
18. Goldfuss, B.; Schleyer, P.V.R.; Hampel, F. A "Lithium-Bonded" Cyclopropyl Edge: The X-ray Crystal Structure of [Li−O−C(Me)−(c-CHCH2CH2)2]6 and Computational Studies. *J. Am. Chem. Soc.* **1996**, *118*, 12183–12189. [CrossRef]
19. Harvey, B.G.; Ernst, R.D. Transition-Metal Complexes with (C−C)→M Agostic Interactions. *Eur. J. Inorg. Chem.* **2017**, 1205–1226. [CrossRef]
20. Sato, F.; Okamoto, S. The Divalent Titanium Complex Ti(O-i-Pr)4/2 i-PrMgX as an Efficient and Practical Reagent for Fine Chemical Synthesis. *Adv. Synth. Catal.* **2001**, *343*, 759–784. [CrossRef]
21. Davis, M.; Deady, L.W.; Finch, A.J.; Smith, J.F. Some reactions of grignard reagents with chloroform and carbon tetrachloride in the presence of cyclohexene. *Tetrahedron* **1973**, *29*, 349–352. [CrossRef]
22. Gartiaa, Y.; Biswasb, A.; Stadlerc, M.; Nasinia, U.B.; Ghosha, A. Cross coupling reactions of multiple CCl bonds of polychlorinated solvents with Grignard reagent using a pincer nickel complex. *J. Mol. Catal. A Chem.* **2012**, *363*, 322–327. [CrossRef]

23. Berding, J.; Lutz, M.; Spek, A.L.; Bouwman, E. Synthesis of Novel Chelating Benzimidazole-Based Carbenes and Their Nickel(II) Complexes: Activity in the Kumada Coupling Reaction. *Organometallics* **2009**, *28*, 1845–1854. [CrossRef]
24. Hayashi, T.; Tajika, M.; Tamao, K.; Kumada, M. High stereoselectivity in asymmetric Grignard cross-coupling catalyzed by nickel complexes of chiral (aminoalkylferrocenyl)phosphines. *J. Am. Chem. Soc.* **1976**, *98*, 3718–3719. [CrossRef]
25. Amruta, J.-P.; Wang, C.-Y.; Biscoe, M.R. Nickel-Catalyzed Kumada Nickel-Catalyzed Kumada Cross-Coupling Reactions of Tertiary Alkylmagnesium Halides and Aryl Bromides/Triflates. *J. Am. Chem. Soc.* **2011**, *133*, 8478–8481. [CrossRef]
26. Kumada, M.; Tamao, K.; Sumitani, K. Phosphine–nickel complex catalyzed cross-coupling of grignard reagents with aryl and alkenyl. *Org. Synth.* **1978**, *58*, 127–133. [CrossRef]
27. Yang, X.; Matsuo, D.; Suzuma, Y.; Fang, J.-K.; Xu, F.; Orita, A.; Otera, J.; Kajiyama, S.; Koumura, N.; Hara, K. Ph2P(O) Group for Protection of Terminal Acetylenes. *Synlett* **2011**, *16*, 2402–2406. [CrossRef]
28. Kondoh, A.; Yorimitsu, H.; Oshima, K. Synthesis of Bulky Phosphines by Rhodium-Catalyzed Formal [2 + 2 + 2] Cycloaddition Reactions of Tethered Diynes with 1-Alkynylphosphine Sulfides. *J. Am. Chem. Soc.* **2007**, *129*, 6996. [CrossRef]
29. Shaibakova, M.G.; Titova, I.G.; Makhmudiyarov, G.A.; Ibragimov, A.G.; Dzhemilev, U.M. Synthesis of 2, 3-acetylenic amines by aminomethylation of acetylenes with geminal diamines. *Russ. J. Org. Chem.* **2010**, *46*, 44–48. [CrossRef]
30. Bieber, L.W.; da Silva, M.F. Mild and efficient synthesis of propargylamines by copper-catalyzed Mannich reaction. *Tetrahedron Lett.* **2004**, *45*, 8281–8283. [CrossRef]
31. Sheldrick, G.M. A Short History of SHELX. *Acta Crystallogr.* **2008**, *64*, 112–122. [CrossRef] [PubMed]
32. Frisch, M.J.; Trucks, G.W.; Schlegel, H.B.; Scuseria, G.E.; Robb, M.A.; Cheeseman, J.R.; Scalmani, G.; Barone, V.; Mennucci, B.; Petersson, G.A.; et al. *Gaussian 09, Revision D.01*; Gaussian Inc.: Wallingford, UK, 2009.
33. Ramazanov, I.R.; Kadikova, R.N.; Saitova, Z.R.; Dzhemilev, U.M. A Route to 1-alkenylphosphine derivatives via the Zr-catalyzed reaction of 1-alkynylphosphines with triethylaluminum. *Asian J. Org. Chem.* **2015**, *4*, 1301–1307. [CrossRef]

© 2019 by the authors. Licensee MDPI, Basel, Switzerland. This article is an open access article distributed under the terms and conditions of the Creative Commons Attribution (CC BY) license (http://creativecommons.org/licenses/by/4.0/).

MDPI
St. Alban-Anlage 66
4052 Basel
Switzerland
Tel. +41 61 683 77 34
Fax +41 61 302 89 18
www.mdpi.com

Catalysts Editorial Office
E-mail: catalysts@mdpi.com
www.mdpi.com/journal/catalysts

www.ingramcontent.com/pod-product-compliance
Lightning Source LLC
LaVergne TN
LVHW070540100526
838202LV00012B/335